PREFACE TO SECOND EDITION

The publication in 1980 of the fifth volume of *Flora Europaea* in which appeared the account of *Carex* by Arthur Chater seemed a very good moment to consider a revision of "British Sedges". In *Flora Europaea* Chater had brought together a certain amount of recent rethinking and opinion combined with his own extensive experience of the genus in Europe and the north temperate hemisphere generally and had arranged the species accordingly; that arrangement is used now in this edition and whilst the species groups will be similar to those found (although not named as formal Sections) in edition one it will be tedious to those who after ten or more years of using the book, will have learnt to accept its order. But such changes are inevitable and need not be excused; I firmly believe it is for the good.

Likewise do I believe that the taxonomy of the two major species complexes (*C. muricata* agg. and *C. flava* agg.) have been revised in a sound and practical way.

That dealing with *Carex muricata*, *C. spicata* and *C. divulsa* is the work of Dick David. The "*flava*" group is a joint revision by both Chater and David, myself acting as devil's advocate or referee when basic concepts were at stake. Again I believe the outcome is an acceptable and sound interpretation of the data as we see it although if any single *Carex* group needs revision on a world basis it is the "Yellow Sedges".

The handbook has been expanded to have a full page drawing of all species described, and the two keys have been restyled with the hope that they are easier to use. A major deviation from other Handbooks is the inclusion of distribution maps based on the records kept at the Biological Records Centre (collected for the most part by members of the B.S.B.I.) and computer-generated under the direction of staff of the Institute of Terrestrial Ecology.

I feel it is not out of place here to say how willingly Arthur Chater and Dick David cooperated with my suggestion they be involved in this revision. Their contribution has been considerable and has undoubtedly raised the standard of this handbook. Furthermore, Dick has taken it upon himself personally to check the field distribution of the rare species and work closely with B.R.C. on the overall map production.

A. C. JERMY

1

We should especially like to thank Dorothy Greene of the Biological Records Centre (I.T.E.) who has devoted many months to ensuring that the data from B.S.B.I. and other sources was punched into the computer and who with colleagues from I.T.E. has produced the 10-km square distribution maps used in this book. Our thanks go also to those B.S.B.I. Recorders who responded so willingly to our request to check these distributions. In particular we are grateful to Gwynn Ellis for allowing us to use his mapping data for *Welsh Flowering Plants* and to Maura Scannell who gave freely her knowledge of the distribution of Irish plants. The additional plates required for this edition were skilfully and painstakingly drawn by Margaret Tebbs. We should like to thank Kathryn Kavanagh for help in preparation of the manuscript.

June 1981.   A. O. CHATER, R. W. DAVID & A. C. JERMY

PREFACE TO FIRST EDITION

We should like to express our thanks to those colleagues who have read and commented upon various parts of the manuscript, in particular to R. W. David, J. Lewis, Dr. A. Melderis, R. Ross and P. D. Sell; and to P. S. Lloyd for his comments on the key to non-fruiting specimens. For their co-operation in obtaining specimens or for data we thank P. M. Benoit, Miss U. K. Duncan, A. Eddy, Miss S. S. Hooper, Miss M. McCallum Webster, A. W. Stelfox, E. L. Swann and the staff of the Biological Records Centre, Monks Wood; we are grateful to Dr. E. Launert for translating Neumann's key. J. E. Dandy made many useful suggestions and allowed us to make use of his MS list of synonyms of British vascular plants. We are especially indebted to Arthur Chater, who has freely given us the benefit of his knowledge of British and European sedges.

A.C.J. & T.G.T.

I should at this point like to acknowledge the help and guidance in my study of *Carex* that I have received over the past years from Professor T. G. Tutin. For me the writing of this book was an exercise from which inevitably I have learnt much, and the part played by Professor Tutin, not only as co-author but also as tutor, is greatly appreciated. My interest in sedges was initiated and encouraged by Dr. J. Heslop-Harrison and it was impossible to go into the field as a student with the late Professor W. H. Pearsall without being made aware of the ecological importance of this group of plants; to both of these teachers I owe a great deal.

The production of this book has been sponsored by the B.S.B.I. and I would especially like to thank J. C. Gardiner, without whose encouragement, assistance and indeed insistence, my contribution would never have been completed. I should lastly like to thank my wife, Alma, for helping with typing, proof-checking and for tolerating over the past year a life dominated by *Carex*.

December 1967. A.C.J.

ABBREVIATIONS

♀	—	female
♂	—	male
☿	—	hermaphrodite
c.	—	*circa*, about
fld, fl.(s)	—	flowered, flower(s)
fr.	—	fruit
infl.	—	inflorescence
lf, lvs	—	leaf, leaves
ppm	—	parts per million
TS	—	transverse section
±	—	more or less

CORRIGENDA

The key on p. 37 has been amended in the text.

p. 70 l. 12   for *infl.* read *spike.*

p. 78 l. 6   *for* 1–4 mm *read* 1–4 cm.

p. 80 l. 10   *for* 3–8 cm *read* 3–8 mm.

p. 98 l. 7   *for* 1–3 mm *read* 1–3 cm.

p. 104 l. 3   *delete comma after* 1–2 mm.

        l. 4   *for* 2–4 mm *read* 2–4 cm.

p. 106 last l. but one *delete C. curta,.*

p. 114 l. 13   *for* 3–10 mm *read* 3–10 cm.

p. 178 l. 12   *insert* mm *after* 4–5.

p. 192 l. 4   *insert* – *between* 1.5 *and* 2.

p. 264   *insert* digitata L. **49** *after* diandra Schrank **3**.

4

CONTENTS

INTRODUCTION

Carex is one of a number of groups of plants which present problems to the non-specialist and was therefore chosen as the subject for what has become the first of this series of Handbooks.

Some groups of *Carex* species can be said to be 'critical', either because of apparent lack of discontinuities between taxa, due to hybridisation, or because vegetative spread has resulted in the production of local biotypes. Although these are distinct enough in minor characters, they embarrass taxonomists by the vastness of their numbers and their slight differentiation from other parts of the population (Heslop-Harrison, 1953). In the latter half of the nineteenth century these biotypes were given undue prominence by the many botanists who were attracted to the genus at a time when every growth-form and minor variant was thought to warrant a name. So the literature is plentifully studded with names at all ranks, from species to sub-forma, and the resulting synonymy complicates the nomenclature. In many Floras produced in Europe at this time the existing plethora of named taxa was added to by the description of regional variants and habitat forms. To add to the confusion, the latter were often given unmerited formal taxonomic rank and have persisted in the literature to this day. At the turn of the century, the increasing number of keen amateur botanists in this country vied with one another in finding new forms; our herbaria are full of their specimens and the literature full of their resulting comments.

In spite of this, and leaving *Carex flava* agg. and the *Carex nigra* group apart, when a wide view of the taxa is taken, *Carex* species in Britain are not too difficult to define. The difficulty often lies, with the beginner at least, in the lack of understanding between himself and the specialist; it is one of terms and definitions not directly comparable to those used for other flowering plants. We hope that this book, and particularly the illustrations, will go some way towards resolving this difficulty. In the following descriptions of the *Carex* species we have emphasized characters which we consider to be influenced by the environment. We have attempted to give a workable key both to fruiting specimens and to vegetative parts alone; the latter includes all broad-leaved Cyperaceae. A synopsis of the groups (Sections) into which the British species fall is given with the hope that this subdivision may aid identification. Ecological groupings of species are appended to act as a guide when dealing with sterile material .The book concludes with a short note on alien species and dubious records and a short discussion on extinct and potentially British species which may be found in the less explored areas of our islands. A full synonymy to British *Carex* species is given in the index.

IDENTIFICATION OF CAREX MATERIAL

It is to be emphasized that specimens can be identified more reliably if fruiting material is available. In ecological studies, however, only vegetative parts may be available, and we have therefore stressed these in the descriptions. The following account of the general structure of the *Carex* plant defines the terms we have used; as far as possible we have used those adopted in previous Floras or works on *Carex* (e.g. Neumann, 1952; Damman, 1963). From it some idea may be gained as to what should be collected or observed in the field in order to name the plant without recourse to exhaustive comparison with herbarium material. The vegetative key refers to fresh material unless otherwise stated, although most characters used can be interpreted from dried herbarium specimens.

Abnormalities which can confuse identification

Occasionally a flower is produced in which the female flower axis will branch at the base of the nut and protrude through the apex of the utricle along with the stigmas. This elongated axis may bear a glume with a further (usually sterile) utricle or a male flower (often bearing fully formed pollen). This throwback to a *Kobresia*-type of inflorescence is further confusing as these aberrant utricles may have three stigmas when it is normal for the species to have two. Such flowers have been found in *C. acuta*, *C. nigra* and *C. flacca*.

Grazing or other damage can stimulate a plant to produce a flowering spike from a basal node which will be atypical of the species. The distribution of the sexes may vary without obvious cause; spikes that are normally male may have female flowers or vice versa. The formation of female spikelets on the site of potential male flowers has been induced experimentally in *C. nigra* and *C. flacca* by D. L. Smith (1967) and this happens also in nature with *C. serotina*, *C. flacca* and *C. pendula*. Male flowers do sometimes appear singly in the middle of an otherwise female spike.

The typical colour of utricles can be masked by a dark pigment produced in certain groups (e.g. *C. nigra* group) usually in strong sunlight and high altitude; exposed plants of *C. saxatilis* also may be highly pigmented. Conditions in nature can affect the colour of rhizome scales and leaf-sheaths and this has been mentioned in the text where relevant.

9

Hybrids

Hybrids, usually between closely related species, are frequent where the parents grow in close proximity; they are usually intermediate morphologically and mostly sterile. Sterility is usually most conveniently judged by examination of the anthers, which remain undehisced and hidden under the glumes; in addition, the nut does not develop in the utricle so that the spike feels soft and is easily flattened between finger and thumb. Most hybrids occur only as isolated plants with the parents, but there are a few hybrids which occur very frequently and which can form extensive populations or uniform swards (e.g. *C. demissa* × *hostiana*, *C. rostrata* × *vesicaria* and *C. elata* × *nigra*).

As Wallace (1975) remarks, *Carex* hybrids have probably been over-recorded in the past; immature plants, or ones flowering or fruiting unsuccessfully because of some inclement (often temporarily unsuitable) factor in the habitat, are often mistaken for hybrids. The identity of a plant that is a hybrid can often better be judged by the field observer, who can see the plant in question in relation to the parents in the same habitat, than by a study of herbarium material.

GENERAL STRUCTURE OF THE CAREX PLANT

The anatomy of *Carex* is covered in detail by Metcalf (1971) where full references to general papers may be found. Previous to this the anatomy of British *Carices* had been studied by Crawford (1910). The following discusses the morphology of the *Carex* plant in relation to the species accounts that follow.

Growth habit: the rhizome and shoot systems

Four kinds of growth habit have been recognized:—

1   A sympodial rhizome system which produces shoots and branches every few nodes, the general direction of growth being upwards, thus forming a tussock (e.g. *C. paniculata*).

2   A sympodial rhizome system in which the internodes between the shoots are few and usually short, radiating from a centre; the shoots, either few or many, form a tufted or caespitose plant (e.g. *C. demissa, C. ovalis*). This group also contains two species (*C. montana, C. divisa*) in which the rhizomes are thick, woody and persistent for some time and may thus be referred to as a (albeit slowly) creeping system. They are often much branched and often form mats rather than tufts; the shoots are always clustered at the growing points. A further variation is seen in *C. digitata* and *C. ornithopoda* where the fertile stem elongates from a lateral shoot and the main apex continues in the vegetative state.

3   A sympodial system in which at least some of the rhizomes are either short- or far-creeping; the shoots are in dense or loose tufts (rarely single shoots, e.g. *C. maritima*) joined by, usually underground, "pioneering" rhizomes, the number of nodes of which may be more or less constant for a given species (e.g. c. 13 in *C. nigra*). This group also contains a few species (e.g. *C. limosa, C. chordorrhiza*) in which the lower internodes of the shoot elongate and become decumbent, often on the surface of the substratum. These species have been called "stoloniferous" although these organs contain food reserves and are technically rhizomes and not stolons. Further shoots arise, singly in the case of *C. chordorrhiza*, from these lower nodes and repeat the pattern.

4   A monopodial rhizome system in which a single or rarely two

11

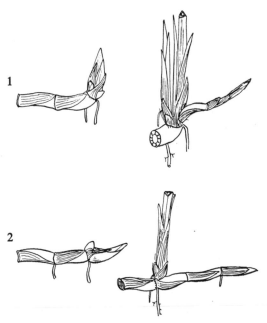

Figures 1–2 Rhizome types. 1 Sympodial; 2 monopodial.

shoots are produced at regular intervals along an often far-creeping, occasionally branched rhizome (e.g. *C. arenaria, C. disticha*).

Thus the creeping, tufted or tussock habit is diagnostic but it can, to some extent, be affected by the habitat. Plants appearing to be tufted may not really be so, as Kershaw (1962) has shown in *C. bigelowii*. Here different "pioneer" rhizomes are "attracted" by each other or are influenced by an outside and hitherto unknown factor (a nutritional gradient?) so that their shoots are produced in a clump. It has been shown experimentally with *C. nigra* that the direction of growth of the rhizomes is towards areas of higher oxygen concentration and away from anaerobic conditions and this may be true for several flush species at least. The importance of this is seen in those *C. nigra* populations growing in wet stagnant conditions, when the pioneer rhizomes will grow upwards to the water or peat surface and will not spread laterally. Eventually these may form a tussock or an island of peat or root mass which may well support a fen flora of its own. Such tussocks have been erroneously quoted in the ecological literature (e.g. Holdgate, 1955, and others) as *C. juncella* (Fries) Th.Fries (*C. nigra* var. *juncea* (Fries) Hyl.) a N Scandinavian taxon. It is imperative then to investigate the true habit of a species and not to be misled by its outward appearance.

Rhizome and root anatomy; rhizome scales; the stem

The roots of peat-loving species tend to be thick, with a small stele and copious surrounding aerenchyma; often a band of sclerenchyma lies beneath the epidermis. Root-hairs are often numerous, slender and form a felt around the root. Secondary roots are sparse in these species. Species of dry or sandy habitats e.g. *C. montana*, *C. maritima*, lack much aerenchyma and have a large stele with the result that the roots are wiry and much branched. The colour of roots from either habitat can be a useful field character but in stagnant (anaerobic) conditions this may be overstained black.

The rhizomes vary in the size of the stele, amount of aerenchyma in the outer cortex and in distribution of sclerenchyma; the epidermis is usually hard and shiny or more rarely corky (woody). There is, as one would expect, a tendency for dry-habitat species to have more bundles or bands of thicker tissue than those of wet areas.

Lanceolate or ovate, keeled leaf-scales arise from each node of the rhizome. They are of a characteristic colour and may be persistent or may decay either completely or partially, leaving only the vascular strands (i.e. become fibrous). The conditions in the substratum must be taken into consideration again when deciding the colour of scales; anaerobic water can produce a false black coloration and acid water can leach out the colour, but usually above the water level signs of the true colour can be seen.

The only aerial stem is that bearing a terminal inflorescence. Usually in *Carex* it is acutely or bluntly trigonous although in a few British species, e.g. *C. dioica*, it is terete and ridged. The illustrations show this outline and also the number of vascular bundles and aerenchyma tubes. Detailed anatomical studies of large samples would be worth making. For the most part, stems are tough and often wiry; in a few species, e.g. *C. aquatilis*, they are brittle and snap on bending even in the fresh state. In some species, mostly in subgenus *Vignea*, long leaf-sheaths form a stem-like shoot called here a false stem.

Leaves, leaf-sheaths and ligules

Carex leaves are on the whole grass-like, varying in their width and the shape of the upper part and apex. In some (e.g. *C. nigra*, *C.*

flacca) the flat, typical leaf structure continues to the tip, the midrib channel gradually petering out. In others (e.g. *C. hostiana*, *C. panicea*) the tip becomes subulate and trigonous in cross-section and the midrib channel ends abruptly below this. Leaves of sterile shoots may not show these characters so clearly as those at the base of the flower-stem. The leaves produced early in the season, i.e. those at the base of the shoot, may be short and generally reduced to almost the dimensions of scales. Most species have flat leaves, or they may be keeled, channelled, inrolled or folded (plicate) (see *Figs 3–7*); this is best seen in transverse section or when a cut leaf is looked at "end-on". A few species (e.g. *C. dioica*) have setaceous leaves which are round or triangular in cross-section. The texture, i.e. softness or rigidity of the leaf tissue, can be diagnostic. Roughness, due usually to forward-pointing fine teeth, is confined to the margins and lower sides of the leaf-veins; these teeth may be lost with age and are best seen on young leaves. Some species exhibit hairiness which is more obvious beneath the leaf than above. This may be reduced or even eliminated by wet conditions of the environment but will always be retained around the leaf-sheath apex or ligule. In one British species, *C. pilulifera*, epidermal papillae are conspicuous (in transverse section) on the upper surface of the leaf; in most species the cuticular pattern present is possibly diagnostic.

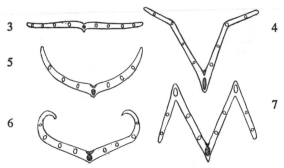

Figures 3–7 Leaf section types. 3 Flat; 4 keeled; 5 channelled; 6 inrolled; 7 plicate.

The anatomy of the leaf has been studied by Crawford (1910) but a thorough investigation over a wide range of material is desirable and might reveal a new range of taxonomic characters. The illustrations later in this book are intended to show nothing more than outline of the transverse section and distribution and number of vascular strands (veins) and tubes of aerenchyma (air spaces). The stomata, which are situated in rows between the veins, could well be investigated; in many species they are present on both surfaces of the leaf but in *C. nigra* for instance they are found only on the *upper* surface. In the closely related *C. acuta* they are confined to the lower epidermis and as Hjelmqvist and Nyholm (1947) have shown leaf anatomy can be very useful in this complex to detect hybrids. The epidermal cells of species in wet habitats are on the whole smaller (e.g. *C. acuta*, 15 μm wide) than those of drier habitats (e.g. *C. hirta*, c. 37 μm). Shepherd (1970, 1972) showed epidermal patterns to be equally specific in species of section *Vesicariae*.

The leaf-sheath (i.e. the lower part of the leaf which is tubular around the shoot or stem) is an important diagnostic part of the plant and must always be observed. The colour, texture and characteristics on decay are important and the remarks about rhizome scales are relevant here, especially in relation to the effect of the habitat. The most important part of the leaf-sheath is the thin, colourless (hyaline) inner face, i.e. that frontal part resulting from the fused leaf-margins. The texture of this is important, e.g. the presence of cross-puckering (as in *C. vulpina*) or brown dots or glands (as in *C. diandra*). The apex of this inner face may be concave, straight or with a tongue-like protruberance (lingulate) (see *Figs 8–10*). This side of the sheath is easily broken on older specimens, becoming V-shaped, and it is best to check younger sheaths for this character. The way in which the sheath splits can be characteristic of a species. Simple tearing is the most common method with the torn hyaline portion soon decaying (e.g. *C. rostrata*) or persisting as a pale brown papery tissue (e.g. *C. acuta*). The other method of splitting produces a ladder-like pattern (i.e. it becomes fibrillose) (see *Fig. 11*); cautious tearing will produce this pattern in those species which exhibit it (e.g. *C. elata*).

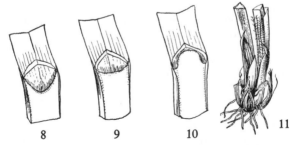

Figures 8–11 Leaf-sheaths; types of inner face apex. 8 Concave; 9 straight; 10 lingulate; 11 sheath showing fibrillae on splitting.

Between the connate leaf-sheath and the flat blade of the leaf is the ligule, as in grasses, but in *Carex* it differs in being fused for most of its length to the upper surface of the blade. If anything, this makes the shape easier to see. The narrow free portion varies in width, is usually hyaline but may in some species be brown or purplish. On the whole the ligule shape reflects the shape of the stem or shoot; in those species where the sterile shoot forms a false stem, the ligule increases in length as the apex of the shoot is approached. The shape is best seen if the leaf is torn back and flattened (see *Figs 12–15*); the length is measured from the apex of the inner face of the sheath to the apex of the ligule.

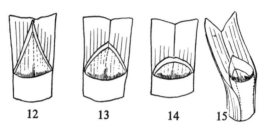

Figures 12–15 Ligule types. 12 Acute; 13 obtuse; 14 rounded; 15 tubular.

The inflorescence; male flowers; female flowers; the *Carex* breeding system

The inflorescence in British *Carex* is basically a spike or a panicle; the flowers (or florets as they are often called) are unisexual but both kinds are borne on the same inflorescence, except in one species, *C. dioica*, which is truly dioecious as the name implies. In subgenus *Carex*, two arrangements of the unisexual flowers are found; the most common is that where the terminal and upper spikes are entirely male and the lower are all female, or occasionally the higher female spikes have male flowers at the top (see *Fig. 16*). A variation on this theme is found in the *C. pulicaris* group where the inflorescence is a single spike with male flowers at the top and female at the base (*Fig. 17*). The other arrangement is seen in Britain only in *C. buxbaumii*, *C. norvegica* and *C. atrata* and here the terminal spike has female flowers at the top and male at the base; the lower spikes may repeat this pattern, or more likely, may be entirely female (see *Fig. 18*). See Smith (1966) and Smith & Faulkner (1976) for account of development.

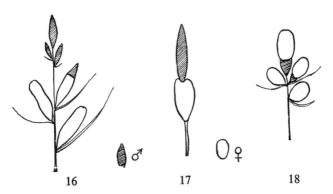

Figures 16–18 Inflorescence types in subgenus *Carex* (see text).

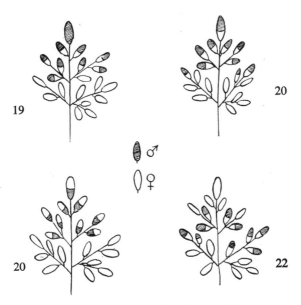

Figures 19–22 Inflorescence types in subgenus *Vignea* (see text).

17

In subgenus *Carex* all spikes except the terminal one are subtended by a bract, which may be like a large glume (not green), or which may be the exaggerated green midrib of the glume otherwise reduced (then called setaceous). On most of the lower spikes at least, however, the green midrib is exaggerated still further to form a leaf-like bract which varies in length according to the species and its position along the axis of the inflorescence. Such bracts may be connate in the lower part and thus ensheath the axis (peduncle) of the spike. That this leafy bract is homologous to the glume subtending the female flower is shown by the presence in the *C. nigra* group of black auricles, remnants of black glume tissue, at the base of the bract; in other species with lighter glumes this is not so evident. The second and more important pointer to this is the very presence of a utricle (albeit sterile and called a *cladoprophyllum*) at the base of the spike axis. Thus it appears that the lateral spike axis in subgenus *Carex* is an elongation of the utricle axis of the more advanced (or reduced) types like *C. pauciflora* or *C. microglochin*; in fact the bristle or *rhachilla* of the latter species is the beginning of that elongation (see *Figs 26, 27*).

The length of the stem over which the spikes are produced, i.e. the length of the inflorescence, varies according to species. In most species this is short in relation to the stem length and the spikes are clustered or contiguous; in some species, however, the internodes between the spikes elongate producing a distant or even remote spike. Occasionally if the apex of the stem is damaged, e.g. grazed, a lower node will produce a remote spike out of character for the species; the bract subtending this spike will be leaf-like and abnormally long, often overtopping the whole inflorescence.

In the subgenus *Vignea* the inflorescence is often more compound (e.g. a panicle) and the spikes (or spikelets as the secondary spikes are often called) are small in comparison with those of many species of subgenus *Carex*. The arrangement of the male and female flowers is basically the same as in subgenus *Carex* and can be one of the following:—(i) The terminal spike male, the lower female with upper spikes male at top (*Fig. 19*). (ii) An elaboration of this where most of the spikes are male at top and female below with a few of the lowest spikes all female (*Fig. 20*). (iii) The terminal and upper spikes female at the top and male below, the lower spikes all female (*Fig. 21*). (iv) The terminal and upper spikes all female, the middle spikes male at the top and female below, and the lower all female again (*Fig. 22*). A variation of this theme is where the terminal spikelet is female, intermediate male and lower female, perhaps with male again at the very base. Bracts in this subgenus are similar to

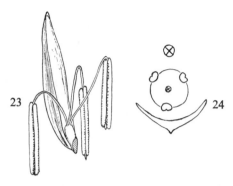

Figures 23–24 Male flower. 23 Adaxial view; 24 floral diagram.

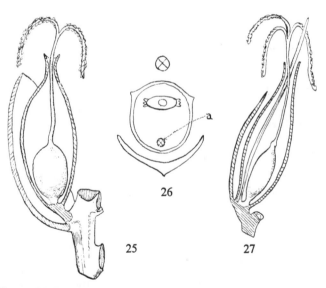

Figures 25–27 Female flower. 25 Longitudinal section; 26 floral diagram (a = position of secondary axis); 27 section of *C. microglochin* flower showing secondary axis (bristle).

those in subgenus *Carex* but on the whole tend to be glumaceous or setaceous; some secondary spikes lack bracts altogether.

The male flowers consist simply of three stamens arising from a low receptacle and subtended on the abaxial side by a glume (see *Fig. 23*). As the basifixed anthers mature the filaments elongate and the stamens hang free from the sheathing glume. Eventually the empty anther falls but the white filament remains thus indicating the position of the male flowers in fruiting specimens. The lower florets of terminal male spikelets may be sterile and have correspondingly larger glumes. The shape of the glume is important but may be affected by the accidental breaking of tissue, giving a false impression of a toothed apex. Colour is usually specific, but in some species (e.g. *C. nigra*) different clones may show a wide range of colour which is not correlated with other characters.

The female flowers consist basically of a bottle-shaped utricle or *perigynium* containing a single ovary and subtended by a glume (*Fig. 25*). The colour, shape and size of the glume are all-important in the identification of a species but, as in the male, laceration or incurling can give a false impression of the apex. The utricle, phylogenetically, is the result of fusion of two inner glumes and therefore exhibits two lateral dominant vascular strands (nerves). The prominence of other nerves to form ribs can be important taxonomically but it must be remembered that on drying, the tissue between the nerves shrinks and accentuates the ribbing. The shape (ranging from obovoid to ovoid to ellipsoid, and usually not changing after fertilisation), size, colour and texture are important taxonomic characters. In some species (e.g. *C. nigra, C. flacca*) a black pigment is produced in utricles exposed to bright sunlight, and plants showing this have wrongly been given varietal or *forma* names. The utricle is usually more elongated at the base of the spikelet and observations should always be made on utricles in a median position. The utricles of several species (e.g. *C. muricata* group) are often inhabited by species of gall flies (e.g. *Wachtliella riparia* Winnertz) and then become abnormally swollen and shiny. The utricle wall may be hairy, papillose or have unicellular spines, especially on the shoulder; a scanning electron microscope examination of wall epidermis patterns has revealed useful taxonomic characters and has been used to show sectional relationships in subgenus *Vignea* by Toivonen & Timonen (1976). The apex of the utricle is drawn out into a beak (rostrum)—the neck of the bottle—which varies in length according to the species and which may be serrate along its lateral edges or smooth; in a few species (e.g. *C. magellanica*) the beak may be absent. The orifice through which the style protrudes may be split, bifid, notched, truncate or obliquely truncate (see *Figs 28–32*).

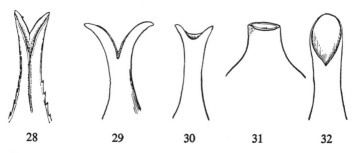

28   29   30   31   32

Figures 28–32 Beak apex types. 28 Split; 29 bifid; 30 notched; 31 truncate; 32 obliquely truncate.

Within the utricle is the ovary which either has two stigmas and is a flat circular disc, maturing into an ellipsoid biconvex nut or has three stigmas and is usually cylindric, maturing into a trigonous, ellipsoid or obovoid nut. In subgenus *Carex* both kinds occur in different species (but see "Abnormalities" on p. 9); in subgenus *Vignea*, as represented in Britain, only the ovary with two stigmas is known. Hybrids between species having 2 and 3 stigmas respectively (e.g. *C.* × *grahamii*) tend to have the two forms in the same inflorescence. The style is straight, or occasionally curved at the base, within the utricle; this difference has been used as a diagnostic character (e.g. in *C. distans*) but is rarely constant; in some species the style is swollen at the point of attachment to the ovary. The style persists in fruit; in some species (e.g. *C. pauciflora*) it may protrude some distance and could be confused with the bristle in *C. microglochin*. The stigmas are papillose and fall soon after fertilization.

The fruit is a biconvex or trigonous nut, which in some species (e.g. *C. montana*) is on a stalk but is usually sessile in the base of the utricle. In some species it fills the utricle upon maturity, in others it does not, giving rise to the condition referred to as an inflated utricle (e.g. *C. rostrata*). The nut is usually brown, sometimes yellowish (*C. distans*) or purple (*C. panicea*); the colour has been used to distinguish species (e.g. *C. distans* and *C. binervis*).

As far as is known species of *Carex* are outbreeding; there is no record of apomixis or vivipary. Protandry, or more rarely protogyny, prevents self-fertilization. Wind is the main pollinating agent although pollen-eating beetles may transport pollen effectively.

21

Most British species are protandrous but those species with several male spikelets (e.g. *C. riparia*) may still be shedding pollen when the first stigmas are receptive lower down the plant. At least in the *C. nigra* group, ripe pollen kept at 40 p.c. humidity for two days can fertilize females lower down the same inflorescence. Certainly in the experimental garden a rammet can be pollinated from inflorescences on rammets of the same clone; this must surely happen in nature too. The stigmas remain receptive usually for 24 hours only; often the whole spikelet is receptive at the same time. The male spikelet on the other hand usually takes one to several days to discharge its pollen, most often in bands of 12 or so florets at a time. Little is known about the time involved in fertilization or even for the formation of a pollen-tube; *Carex* pollen is difficult to germinate on artificial media.

Chromosome studies

Chromosome studies on British and European *Carex* in the past 25 years have been few. Davies (1956) discusses the origin of chromosome numbers and gives counts for many British species. More recently Faulkner (1972) reports on his work on the section *Acutae* and gives a good review of the cytology of *Carex* generally. One significant point emerges from his work, namely that on cytological grounds Scottish *C. recta* appears to be an interspecific hybrid between *C. aquatilis* and the hitherto non-British *C. paleacea*. Studies on hybrids within the *Acutae* suggest there is considerable homology between chromosomes of the different species. A similar situation may well exist in the *C. flava* group.

Chromosome numbers of *Carex* exhibit aneuploid series. The highest number recorded for Britain is that of *C. hirta* (2n = 112); the lowest *C. panicea* and *C. vaginata* (2n = 32).

Chemistry of *Carex*

Harborne (1971) has investigated the flavonoids in Cyperaceae. Toivonen (1974) looked at members of the *Canescentes* (as *Heleonastes*) section using paper chromatography and found flavonoid patterns useful. In spite of their morphological similarity *C. lachenalii* and *C. glareosa* (see p. 251) apparently have different chromatograms. There is obviously scope for more studies of the chemistry of *Carex* species, especially in relation to sectional classification.

ECOLOGY OF CAREX

The majority of *Carex* species are, by reason of their habit and mode of vegetative growth, important if not dominant in a wide variety of plant communities. It is therefore surprising that more autecological accounts are not available; *C. flacca* is the only species of which a Biological Flora account (Taylor, 1953) is available at the time of writing, although several are in the course of being prepared. Notes on the ecology of British species may be found scattered through the pages of the *Botanical Exchange Club Reports* and of the *Proceedings of the B.S.B.I.* (see Simpson, 1960). Specific recent papers of significance include Bernard, 1976; Callaghan, 1976; Coombe, 1954; David, 1977, 1978a, 1978b, 1979a, 1979b, 1980a, 1980b, 1981a, 1981b, 1982; Fitter & Smith, 1979; Mitchell & Stirling, 1980; Noble *et al.*, 1979; Wells, 1975.

Seed dispersal; germination; autecological studies

It is interesting to note that in two riparian species (*C. elata* and *C. pseudocyperus*), which can only be effectively established in a very narrow band along the side of a slow-flowing stream or dyke, utricles are dropped singly on maturity and so decrease the competition factor that would occur if the whole spikelet fell into the water. There are few special means of seed dispersal in *Carex*. Flotation and water dispersal are obvious in aquatic species, especially in those with inflated utricles (e.g. *C. vesicaria*, *C. rostrata*). According to Wiinstedt (1945) the papillose nature of the utricle wall in *C. flacca* enhances buoyancy. Wind and the possible carrying of utricles by ants and other foraging insects can disperse terrestrial species short distances only. On the whole little is known of the dispersal, germination and establishment of *Carex* species in Britain and both observation and experimentation offer scope for interesting work.

Most *Carex* species require an after-ripening period of 3–12 months; low-temperature treatment does not substantially reduce this. Artificial abrasion of the testa increases the percentage germination; complete removal of the utricle (in *C. nigra*, at least) produces a lower percentage, which suggests some substance there affecting imbibition rate. In *C. flacca* a colourless mucilage is exuded from the nut before the testa breaks but this has not been observed in the germination of aquatic species. Light certainly hastens germination; *C. nigra* sown under soil in pots took six months before sprouting, whereas when sown on damp filter-paper in the light it germinated within 14 days.

Autecological studies of dominant *Carex* species will undoubtedly show some species to be good ecological indicators especially in relation to pH, calcium, phosphate, carbon/nitrogen ratio and potentially toxic ions as manganese, iron and aluminium. Moving water of flushes gives aeration, necessary for good rhizome growth in some species, and a constant supply of nutrients. The delicately balanced, but slightly different, ecological requirements of closely related species can be useful in identification. A detailed analysis of these requirements, on the other hand, can show the species, once identified, to be a good ecological pointer. Such differences in species groups can prevent competition for available space. The effect of aluminium in relation to calcium has been demonstrated in *C. lepidocarpa* and *C. demissa* by Clymo (1962). A positive correlation between soil moisture and plant size in *C. aquatilis* was shown by Shaver *et. al.* (1979).

The importance of *Carex* in the primary hydrosere is well established, e.g. *C. rostrata* as a colonist of open acid water of upland lakes. The position of *C. paniculata* in the premature establishment of alder ('swamp') carr in the Norfolk Broads and its later stabilization by *C. acutiformis* and *C. riparia* has been shown by Lambert (1951). The composition of *Carex*-dominated flushes in the N Pennines has been described by Eddy & Welch (1969).

Habitat definitions of *Carex* dominated vegetation

The terms oligotrophic (base-poor and generally acidic) and eutrophic (rich in base cations and neutral or alkaline in reaction) are used to amplify the ecological notes. Mesotrophic is an intermediate condition. Eutrophic conditions (i.e. added cations) are imposed upon oligotrophic mires in lowland coastal situations by deposition from salt-laden winds and sea-spray (Holden, 1961; Jermy, Hibberd & Sims, 1978).

Peatlands of which Cyperaceae are a major constituent may be classified either according to their different physical (topographical, hydrological and chemical) characteristics or by their present composition of vegetation (i.e. species), or by a combination of the two. Nomenclature of peatlands in Britain (blanket bogs, fens, mires, swamps, valley bogs etc.) has become a little confused due to loose usage and we use here the definitions proposed by Ratcliffe (1977) for the *Nature Conservation Review* as being more meaningful in terms of *Carex* ecology. Plant communities are referred to in this book only where their components are distinctive, e.g. *Sphagnum* communities, although a number of major accounts describe *Carex*-dominated vegetation on this basis (Birse & Robertson, 1967, 1973, 1976; McVean & Ratcliffe, 1962; Ratcliffe, 1964a, b; Spence, 1964; Wheeler, 1978, 1980a, b, c.).

Thus in the ecological notes for each species we have used the following terminology (after Ratcliffe, 1977) for wet peatlands in all of which *Carex* predominates:

A. *Ombrogenous mires* (in which the water component originates through rainfall).

   i. *blanket mires* (blanket bogs)—continuous acidic areas of peat-forming vegetation over flat or sloping ground with impeded drainage, *Sphagnum* dominated and mainly upland.

   ii. *raised mires* (raised bogs)—usually deep (up to 10m) peat often formed in initially more eutrophic conditions; *Sphagnum* is a dominant component.

B. *Topogenous mires* (in which the local relief results through drainage in a permanently high water table).

   iii. *open water transition* (reed-swamp) and *flood plain mires* (fens)—usually meso- or eutrophic mires in E and S Britain in which oligotrophic nuclei may develop locally and accommodate acid-loving or base-poor species. Flood plain mires on the siliceous rocks of N and W Britain are usually more oligotrophic. Open water mires grade or develop into flood plain mires and both contain seral communities.

   iv. *basin mires*—develop in enclosed waterlogged depressions and frequent in areas of glacial deposition, in which ground water plays a significant part in the overall mire water budget. These mires are similar to raised mires in their oligotrophic *Sphagnum* communities although 'brown' (Hypnoid) mosses are more frequent and *Carex lasiocarpa* is an indicator of mesotrophic conditions.

   v. *valley mires* (valley bogs)—occur in small, shallow valleys or channels in which there is lateral water movement. They often show a range of base content and correlated vegetation mosaic and are often very similar to ombrogenous raised mires and soligenous mires.

C. *Soligenous mires* (flushes and spring communities). Associated with surface water seepage of varying chemical content on slopes of all kinds. Characteristically these often base-rich areas present communities richer in species than the surrounding and often base-poor ombrogenous mires. They are frequently grazed, especially in upland areas, and the resulting community is low in stature. Different *Carex* species are indicators of different content of base cations but this is not fully understood and there is scope for study in the requirements of soligenous mire Cyperaceae.

The relationships between Cyperaceae associated with soligenous mires (flushes) and the chemical (base cations) content of the ground water are tentatively shown in the following table:

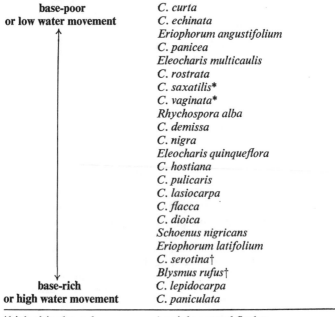

base-poor or low water movement	*C. curta*
	C. echinata
	Eriophorum angustifolium
	C. panicea
	Eleocharis multicaulis
	C. rostrata
	C. saxatilis*
	C. vaginata*
	Rhychospora alba
	C. demissa
	C. nigra
	Eleocharis quinqueflora
	C. hostiana
	C. pulicaris
	C. lasiocarpa
	C. flacca
	C. dioica
	Schoenus nigricans
	Eriophorum latifolium
	C. serotina†
	Blysmus rufus†
base-rich or high water movement	*C. lepidocarpa*
	C. paniculata

*high altitudes only   †mainly coastal flushes

Habitat lists (rare and local species indicated by *)

Stream, dyke and pond edges

C. acuta, acutiformis, aquatilis, elata, elongata*, otrubae, paniculata, pseudocyperus, riparia, vesicaria, vulpina*.

Base-rich mires (usually on deep organic soil)

C. acutiformis, appropinquata, buxbaumii*, diandra, disticha, elata, flava*, lasiocarpa, lepidocarpa, nigra, paniculata, riparia, serotina, vesicaria.

Base-poor mires (including valley mires on deep organic soil)
C. chordorrhiza*, curta, demissa, dioica, echinata, elata, hostiana, lasiocarpa, limosa, nigra, panicea, paniculata, pulicaris, rostrata, serotina, vesicaria.

Base-rich flushes (soligenous mires on shallow organic soils on steep slopes)
C. dioica, flacca, hostiana, lepidocarpa, nigra, pulicaris, rupestris*.

Base-poor flushes (shallow organic soils on steep slopes)
C. curta, demissa, dioica, echinata, hostiana, lachenalii*, panicea, rariflora*, rostrata, saxatilis.

Oligotrophic (i.e. Sphagnum) ombrogenous mires
C. curta, echinata, limosa, nigra, pauciflora, magellanica, pulicaris, rostrata.

Wet corries and ledges above 600 m (2000 ft) altitude
C. atrata, atrofusca*, bigelowii, binervis, flacca, lachenalii*, microglochin*, nigra, norvegica*, panicea, pulicaris, rupestris*, saxatilis, vaginata.

Mountain acid grassland
C. bigelowii, binervis, hirta, nigra, ovalis, pilulifera.

Chalk/limestone grassland
C. capillaris, caryophyllea, ericetorum, tomentosa*, flacca, humilis*, montana, ornithopoda*.

Dry limestone woodland/scrub
C. depauperata*, digitata*, muricata *subsp.* muricata*, tomentosa*, sylvatica.

Sand-dunes, heaths and damp acid grassland
C. arenaria, binervis, flacca, hirta, nigra, ovalis, panicea, pilulifera, serotina.

Roadsides, hedgerows, rough pasture/scrub
C. divulsa, tomentosa*, hirta, muricata *subsp.* lamprocarpa, otrubae, ovalis, pilulifera, spicata, sylvatica.

Lowland wet or clay woodland
C. elongata*, laevigata, nigra, paniculata, pendula, remota, strigosa, sylvatica, vesicaria.

Sea-spray zone, brackish ditches, estuarine flats
C. distans, divisa, extensa, maritima*, nigra, otrubae, punctata, recta*, serotina.

CLASSIFICATION

The Cyperaceae show a considerable diversity in both their vegetative and reproductive parts. The classification of the major subdivisions of the family has been based on a combination of the sex of the flowers, the numbers of fertile flowers within a spikelet and the shape of fused inner glumes called the prophyll (see below). A general synopsis as it relates to *Carex* is shown here.

Family: Cyperaceae

Subfamily: 1 Mapanioideae
2 Scirpoideae
3 Rhynchosporoideae
4 Caricoideae:

Tribe: Cariceae: (the only one in the subfamily)

Genus: 1 Hoppia Nees (S American)
2 Schoenoxiphium Nees (African)
3 Kobresia Willd. (N Temperate, mainly Himalayan)
4 Uncinia Pers. (Australasian, S American)
5 Carex L. (widespread):

Subgenus: 1 Vignea (Beauv. ex Lestib.) Kük.
2 Carex
3 Primocarex Kük.

The *Cariceae* are characterised by having unisexual flowers which lack a perianth and by the pistillate (i.e. female) flowers being ensheathed by the inner glume or prophyll. The distinctions and generic limits, often debated, between *Hoppia*, *Schoenoxiphium* and *Uncinia* need not be discussed here. *Kobresia* on the other hand, will be more familiar—*K. simpliciuscula* (Wahlenb.) Mackenzie is found in Britain and *K. myosuroides* (Vill.) Fiori and Paol. in Europe; it differs from *Carex* in having open, spathe-like prophylls and from 1 to 6 male flowers on the same axis as the female flower and subtended by the same bract. *Carex*, contrastingly, has a closed, connate (bottle-like), prophyll, called the utricle, which has only one female flower within it and no male flowers arising on the same axis (but see p. 9).

It may not be out of place here, to point out the main differences between the Cyperaceae (and particularly *Carex*) and the Gramineae; both are often rhizomatous herbs with linear leaves, some of which may be reduced to sheaths, and with reduced flowers composed of glumes and/or bristles grouped together in spikes or spikelets. The following table contrasts the two groups:

Sedges	Grasses
1. *Stem* usually trigonous with an extended internode just below the inflorescence.	*Stem* terete, with nodes evenly spaced throughout.
2. *Leaf-sheath* never split.	*Leaf-sheath* usually split.
3. *Ligule* partially fused to upper surface of leaf-blade.	*Ligule* not fused to leaf-blade.
4. *Bracts* subtending spikes or spikelets often leaf-like.	*Bracts* subtending spikes or spikelets never leaf-like.
5. *Flowers* spirally arranged and appearing in 4 or 5 rows (except *Cyperus* where they are distichous).	*Flowers* and spikelets distichously arranged
6. Anthers basifixed.	Anthers dorsifixed.

Subgenera and species grouping in British *Carex*

The genus *Carex* contains over 1800 species throughout the world although the majority are N Temperate; in Europe alone there are 180 (Chater, 1980). Kükenthal (1909) in his monograph on the genus in Engler's *Das Pflanzenreich* series, recognised four subgenera: *Primocarex*, *Vignea*, *Indocarex* and *Eucarex*. Kreczetowicz (1936) queried the primitiveness of the species in subgenus *Primocarex* and instead placed them mainly into subgenus *Carex* with a few in subgenus *Vignea*. Kern and Reichgelt (1954) in their work on the Dutch Carices likewise dispensed with *Primocarex*, as did others (e.g. Koyama, 1962). On the other hand Saville and Calder (1953), basing their argument on evidence from smut-fungi-*Carex*-host relationship proposed a further subgenus, *Kuekenthalia* (in honour of Georg Kükenthal) and in it included the predominantly wetland and aquatic species with well developed, inflated utricles, which have developed a seed-dispersal mechanism. They included in it some species, e.g. *C. pulicaris* and *C. pauciflora*, included by Kükenthal in subgenus *Primocarex* and Kern in *Vignea*. In edition 1 of this handbook Jermy & Tutin followed Kern and recognized two subgenera only, *Carex* and *Vignea*.

The classification adopted in this second edition follows that of *Flora Europaea* Vol. 5 (1980). This classification is a compromise, in that subgenus *Primocarex* has been invoked as a convenient heading under which certain anomalous species can be placed, although this subgenus hardly seems to have any other justification for its existence and cannot be considered to have any phyletic significance. A synopsis of the classification is given here as an understanding of these groupings and some appreciation of the logic behind the ordering of the species can be a great help in mastering the genus as a whole.

The major division in this classification is into two subgenera. Subgenus *Vignea* is characterised most obviously by all the spikes being similar in appearance, as in *C. ovalis* or *C. paniculata*; there are always only two stigmas. Within this subgenus, the sections are characterised chiefly by the distribution of male and female flowers within the spikes. *C. dioica* and *C. davalliana* are the only dioecious species with a solitary spike. Unfortunately in the British Isles the sections of subgenus *Vignea* never contain more than three species, so they are not as helpful as they are on a European scale. In subgenus *Carex*, however, several sections are larger and are useful for the beginner to learn to recognise. Subgenus *Carex* is especially characterised by having the terminal one or more male spikes very different in appearance from the lateral, female ones, as in *C. pendula* and *C. nigra*. Most of the species have three stigmas, the only ones with two stigmas being the *C. saxatilis* and the *C. nigra* group which are among the most difficult for the beginner. Species with hairy utricles fall into four sections: *C. hirta* and *C. lasiocarpa* are the only two of these species with a prominent 2-fid beak on the utricle; of the rest, *C. digitata*, *C. ornithopoda* and *C. humilis* are the only ones with long, narrow female spikes; *C. caryophyllea* has ovoid spikes and the lowest bract sheathing, while *C. tomentosa*, *C. ericetorum*, *C. montana* and *C. pilulifera* have ovoid spikes and the lowest bract not sheathing. *C. atrata* and its two allies, *C. buxbaumii* and *C. norvegica*, look at first sight as though they should belong to subgenus *Vignea* as the spikes all look the same, but the presence of three stigmas makes it clear that their correct position is as a section of subgenus *Carex*. Two common and closely related sections can cause especial confusion: the section containing *C. distans*, *C. punctata*, *C. binervis* and *C. extensa* can be distinguished from that containing *C. hostiana* and the *C. flava* group chiefly by the usually mucronate female glumes and speckled utricles; the latter group has obtuse to subacute female glumes and unspeckled utricles. Monoecious species of *Carex* with a solitary spike will all be found in the highly artificial but convenient subgenus *Primocarex*. From comparative studies of inflorescence development Smith &

Faulkner (1976) conclude that members of this subgenus, although possibly polyphyletic in origin, are highly advanced.

Synopsis of classification and arrangement of species

Subgen. VIGNEA

Sect. Heleoglochin Dumort.
1 paniculata L.
2 appropinquata Schumacher
3 diandra Schrank

Sect. Vulpinae (J. Carey) Christ
4 vulpina L.
5 otrubae Podp.

Sect. Phaestoglochin Dumort.
6 spicata Hudson
7 muricata L.
   a. subsp. muricata
   b. subsp. lamprocarpa Čelak.
8 divulsa Stokes
   a. subsp. divulsa
   b. subsp. leersii (Kneucker) Walo Koch

Sect. Ammoglochin Dumort.
9 arenaria L.
10 disticha Hudson

Sect. Divisae Christ ex Kük.
11 chordorrhiza L. fil.
12 divisa Hudson

Sect. Foetidae (L. H. Bailey) Kük.
13 maritima Gunnerus

Sect. Remotae (Ascherson) C. B. Clarke
14 remota L.

Sect. Ovales (Kunth) Christ
15 ovalis Good.

Sect. Stellulatae (Kunth) Christ
16 echinata Murray

Sect. Physoglochin Dumort.
17 dioica L.

Sect. Elongatae (Kunth) Kük.
18 elongata L.

Sect. Canescentes (Fries) Christ
19 lachenalii Schkuhr
20 curta Good.

31

Subgen. CAREX

Sect. Carex
21 hirta L.
22 lasiocarpa Ehrh.

Sect. Paludosae (Fries) Christ
23 acutiformis Ehrh.
24 riparia Curtis

Sect. Pseudocypereae (L. H. Bailey) Kük.
25 pseudocyperus L.

Sect. Vesicariae (O. F. Lang) Christ
26 rostrata Stokes
27 vesicaria L.
28 saxatilis L.
29 ×grahamii Boott

Sect. Rhynchocystis Dumort.
30 pendula Hudson

Sect. Strigosae (Fries) Christ
31 sylvatica Hudson
32 capillaris L.
33 strigosa Hudson

Sect. Glaucae (Ascherson) Christ
34 flacca Schreber

Sect. Paniceae (O. F. Lang) Christ
35 panicea L.
36 vaginata Tausch

Sect. Rhomboidales Kük.
37 depauperata Curtis ex With.

Sect. Elatae Kük.
38 laevigata Sm.

Sect. Spirostachyae (Drejer) E. H. Bailey
39 binervis Sm.
40 distans L.
41 punctata Gaudin
42 extensa Good.

Sect. Ceratocystis Dumort.
43 hostiana DC.
44 flava L.
45 lepidocarpa Tausch
46 demissa Hornem.
47 serotina Mérat

Sect. Porocystis Dumort.
48 pallescens L.

Sect. Digitatae (Fries) Christ
    49 digitata L.
    50 ornithopoda Willd.
    51 humilis Leysser
Sect. Mitratae Kük.
    52 caryophyllea Latourr.
Sect. Acrocystis Dumort.
    53 tomentosa L.
    54 ericetorum Pollich
    55 montana L.
    56 pilulifera L.
Sect. Aulocystis Dumort.
    57 atrofusca Schkuhr
Sect. Limosae (O. F. Lang) Christ
    58 limosa L.
    59 rariflora (Wahlenb.) Sm.
    60 magellanica Lam.
          subsp. irrigua (Wahlenb.) Hiitonen
Sect. Atratae Christ
    61 atrata L.
    62 buxbaumii Wahlenb.
    63 norvegica Retz.
Sect. Phacocystis Dumort.
    64 recta Boott
    65 aquatilis Wahlenb.
    66 bigelowii Torrey ex Schweinitz
    67 elata All.
    68 nigra (L.) Reichard
    69 acuta L.

Subgen. PRIMOCAREX

Sect. Leucoglochin Dumort.
    70 microglochin Wahlenb.
    71 pauciflora Lightf.
Sect. Petraeae (O. F. Lang) Kük.
    72 rupestris All.
Sect. Unciniformes Kük.
    73 pulicaris L.

KEY TO CAREX SPECIES IN FRUIT

1a	Spike solitary	2
1b	Spikes 2 or more	6
2a (1)	Stigmas 2; nut biconvex	3
2b	Stigmas 3; nut trigonous	4
3a (2)	Dioecious; utricles 2.5–3.5 mm	**17 dioica**
3b	Monoecious; utricles 4–6 mm	**73 pulicaris**
4a (2)	Utricles erecto-patent to erect at maturity; ♀ glumes persistent; leaves ± curled at apex	**72 rupestris**
4b	Utricles deflexed at maturity; ♀ glumes caducous; leaves ± straight	5
5a (4)	Utricles 3.5–4.5 (−6) mm, with a bristle arising from base of nut and protruding from the top of the beak along with the style-base	**70 microglochin**
5b	Utricles 5–7 mm, without a bristle ..	**71 pauciflora**
6a (1)	Spikes all ± similar in appearance, the terminal usually at least partly ♀	7
6b	Spikes dissimilar in appearance, the terminal or upper usually entirely ♂, the lower usually entirely ♀	31
7a (6)	Stigmas 2; nut biconvex	8
7b	Stigmas 3; nut trigonous	28
8a (7)	All spikes with ♀ fls at apex	9
8b	At least one spike with ♂ fls at apex ..	14
9a (8)	Body of utricle distinctly winged for at least part of its length	**15 ovalis**
9b	Utricles unwinged, except sometimes narrowly so on the beak	10
10a (9)	Lowest bract lf-like, exceeding infl. ..	**14 remota**
10b	Lowest bract usually not lf-like, shorter than infl.	11
11a (10)	Spikes subglobose; utricles usually very divaricate at maturity, not more than 10 ..	**16 echinata**
11b	Spikes ovoid to oblong; utricles erect or erecto-patent, more than 10	12
12a (11)	Spikes whitish, greenish or pale brown ..	**20 curta**

35

12b	Spikes dark reddish-brown	13
13a (12)	Spikes (5–) 8–12 (–18); utricles 3–4 mm, without a slit down the back of the beak; robust plant of wet woods and ditches ..	**18 elongata**
13b	Spikes (2–) 3–4 (–5); utricles (1.5–) 2.5–3 mm, with a slit down the back of the beak; high alpine	**19 lachenalii**
14a (8)	Plant densely caespitose, without creeping rhizomes	15
14b	Plant not or laxly caespitose, with creeping rhizomes	24
15a (14)	Utricles weakly to strongly convex on adaxial side, strongly convex on abaxial side	16
15b	Utricles plane on adaxial side, weakly convex on abaxial side	18
16a (15)	Utricles broadly winged in upper half ..	**1 paniculata**
16b	Utricles not or only very narrowly winged	17
17a (16)	Usually tussock-forming; basal sheaths fibrous; lower clusters of spikes ± pedunculate	**2 appropinquata**
17b	Not tussock-forming; basal sheaths entire; lower spikes or clusters of spikes sessile ..	**3 diandra**
18a (15)	Stems more than 2 mm wide; leaves (2–) 4–8 (–10) mm wide; utricles distinctly nerved ± throughout	19
18b	Stems less than 2 mm wide; leaves 2–4 (–5) mm wide; utricles nerveless except for faint nerves at the base	20
19a (18)	Ligule longer than wide, not overlapping edges of lf; utricles shining, smooth, with oblong, thin-walled epidermal cells ..	**5 otrubae**
19b	Ligule much wider than long, overlapping edges of leaf; utricles dull, papillose, with isodiametric, thick-walled epidermal cells	**4 vulpina**
20a (28)	Roots, and often basal sheaths and base of stems, ligules and occasionally glumes purplish-tinged; ligule distinctly longer than wide; utricles corky and thickened at base	**6 spicata**
20b	Roots, basal sheaths and base of stems not purplish-tinged; ligule not or only slightly longer than wide; utricles not corky and thickened at base	21

21a (20)	Lowest 3–4 spikes (or branches) separated from each other by a gap of much more than their own length; ripe utricles not divaricate	**8a divulsa** subsp. **divulsa**
21b	Lowest spikes overlapping, or separated from each other by a gap of not more than their own length; ripe utricles divaricate ..	22
22a (21)	Ligule usually wider than long; infl. 3–5 (–8) cm; utricles at least 4.5 mm long, $\pm$ equally narrowed at both ends ...	**8b divulsa** subsp. **leersii**
22b	Ligule about as wide as long; infl. 2–3 (–4) cm; urticles less than 4.5 mm long, truncate or rounded at base	23
23a (22)	Spikes globose; ♀ glumes blackish—or dark reddish-brown, much darker and shorter than the greenish or brownish utricles; utricles (3.5–)4–4.5 mm, strongly divaricate	**7a muricata** subsp. **muricata**
23b	Spikes ovoid; ♀ glumes pale brown, similar in colour to or paler than, and almost as long as the utricles; utricles 3–3.5 mm, erecto-patent	**7b muricata** subsp. **lamprocarpa**
24a (14)	Body of utricle distinctly winged for at least part of its length	24A
24b	Utricle unwinged, except sometimes narrowly so on the beak	25
24Aa (24)	Middle spikes entirely ♂; terminal or upper spikes entirely ♀	**10 disticha**
24Ab	Middle spikes ♂ at top, ♀ below; terminal spike entirely ♂	**9 arenaria**
25a (24)	Stems smooth	26
25b	Stems rough at least towards the top ..	27
26a (25)	Stems 1–18 cm; utricles gradually narrowed into beak; coastal sands or rocks	**13 maritima**
26b	Stems 15–40 cm; utricles abruptly contracted into beak; very wet bogs ..	**11 chordorrhiza**
27a (25)	Laxly caespitose, with short, ascending rhizomes; beak comprising more than $\frac{1}{3}$ of the length of the utricle; lowest bract shorter than its spike	**3 diandra**

27b	Rhizomes far-creeping; beak comprising less than $\frac{1}{3}$ of the length of the utricle; lowest bract longer than its spike and usually longer than the whole infl. ..	**12 divisa**
28a (7)	Lowest spike erect	29
28b	Lowest spike drooping or pendent ..	30
29a (28)	Spikes in a compact cluster; utricles greenish-brown; high alpine ledges and flushes	**63 norvegica**
29b	Spikes ± remote or the lowest arising at least 1 cm from the one above; utricles pale green; lakeside mires	**62 buxbaumii**
30a (28)	Lowest bract with a sheath at least 5 mm; terminal spike ♂ at top, ♀ at base	**57 atrofusca**
30b	Lowest bract not sheathing or with a sheath less than 3 mm; terminal spike ♀ at top, ♂ at base	**61 atrata**
31a (6)	Stigmas 2; utricles usually plano-convex or biconvex; nuts biconvex	32
31b	Stigmas 3; utricles usually trigonous or inflated; nuts trigonous	40
32a (31)	Utricles inflated, patent	33
32b	Utricles not inflated, erect or erecto-patent	34
33a (32)	Utricles 3–3.5 mm, ovoid, fertile	**28 saxatilis**
33b	Utricles at least 4 mm, ovoid-ellipsoid, sterile	**29 × grahamii**
34a (32)	Densely caespitose, sometimes tussock-forming, without creeping rhizomes ..	35
34b	Not or laxly caespitose, with creeping rhizomes	36
35a (34)	Basal sheaths brown or reddish brown, not fibrous; margins of lvs rolling inwards on drying, the stomata mostly confined to the upper surface; lowest bract usually exceeding spike but not infl.	**68 nigra**
35b	Basal sheaths yellowish-brown, becoming conspicuously reticulately fibrous; margins of lvs rolling outwards on drying, the stomata confined to the lower surface; lowest bract usually shorter than spike ..	**67 elata**
36a (34)	Utricles nerveless	37
36b	Utricles nerved (sometimes obscurely so)	38

37a (36)	Lowest bract exceeding infl.; stems bluntly trigonous, brittle	**65 aquatilis**
37b	Lowest bract not exceeding infl.; stems sharply trigonous, not brittle	**66 bigelowii**
38a (36)	♀ glumes, at least in lower part of spikes, aristate, up to 5 times as long as utricles ..	**64 recta**
38b	♀ glumes obtuse or acute, not more than 1½ times as long as utricles	**39**
39a (38)	Lvs 1–3(–5) mm wide, the margins rolling inwards on drying, the stomata mostly confined to the upper surface; lowest bract narrow, not exceeding or concealing infl.; ♂ spike usually solitary	**68 nigra**
39b	Lvs 3–10 mm wide, the margins rolling outwards on drying, the stomata confined to the lower surface; lowest bract exceeding and often concealing infl.; ♂ spikes usually 2–4	**69 acuta**
40a (31)	Utricles hairy on at least part of the surface of the body	**41**
40b	Utricles glabrous on surface of body, though sometimes ciliate or hispid-denticulate on margin or on beak	**51**
41a (40)	Utricles with a prominent, bifid beak usually more than 0.5 mm	**42**
41b	Utricles without or with a short, usually conical, entire or notched beak not more than 0.5 mm	**43**
42a (41)	Utricles 3.5–5 mm; leaves less than 2 mm wide, glabrous	**22 lasiocarpa**
42b	Utricles 5–7 mm; leaves more than 3 mm wide, hairy at least on the sheaths	**21 hirta**
43a (41)	Plant not or only laxly caespitose, with creeping rhizomes	**44**
43b	Plant ± densely caespitose, without creeping rhizomes	**47**
44a (43)	♂ spikes usually 2 or more; utricles papillose; basal sheaths remaining entire	**34 flacca**
44b	♂ spike solitary; utricles strongly hairy; basal sheaths becoming fibrous	**45**
45a (44)	Basal sheaths red, shiny; stems usually more than 20 cm; leaves erect; lowest bract exceeding spike	**53 tomentosa**

45b	Basal sheaths brown; stems usually less than 20 cm; leaves ±recurved; lowest bract usually shorter than spike	46
46a (45)	Lowest bract with a sheath 3–5 mm; ♀ glumes equalling utricles, acute, green or brownish, without or with a narrow scarious margin	**52 caryophyllea**
46b	Lowest bract not sheathing or with a sheath less than 2 mm; ♀ glumes usually shorter than utricles, obtuse, purplish-black, with a wide scarious and often ciliate margin	**54 ericetorum**
47a (43)	Infl. comprising ± all of stem; ♀ spikes with 2–4 fls.	**51 humilis**
47b	Infl. comprising not more than $\frac{2}{3}$ of stem; ♀ spikes with usually more than 4 fls ..	48
48a (47)	♀ spikes not more than 3 mm wide, lax; flowering stems lateral, leafless	49
48b	♀ spikes 4–6 mm wide, dense; flowering stems terminal, leafy at base	50
49a (48)	Utricles 3–4.5mm, ± equalling purplish glumes; ♀ spikes separated	**49 digitata**
49b	Utricles 2–3 mm, $1\frac{1}{2}$–2 times as long as straw-coloured glumes; ♀ spikes all arising from ± the same point	**50 ornithopoda**
50a (48)	Utricles 3–4.5 mm, pyriform; leaves soft, pale green; basal sheaths reddish-brown ..	**55 montana**
50b	Utricles 2–3.5 mm, obovoid; leaves ± rigid, mid- or grey-green; basal sheaths brown	**56 pilulifera**
51a (40)	At least the lowest spike pendent	52
51b	Spikes not pendent	68
52a (51)	♂ spikes 2 or more	53
52b	♂ spike solitary	58
53a (52)	Utricle without or with an entire or weakly notched beak less than 0.5 mm ..	54
53b	Utricles with a usually strongly bifid beak more than 0.5 mm	55
54a (53)	Leaves more than 7 mm wide; lowest bract with a sheath (30–)50–100 mm	**30 pendula**
54b	Leaves less than 7 mm wide; lowest bract not sheathing or with a sheath not more than 3(–10) mm	**34 flacca**

66a (65)	♀ spikes 3–5 mm wide; utricles with smooth beak	**31 sylvatica**
66b	♀ spikes 5–8 mm wide; utricles with scabrid beak	67
67a (66)	Leaves 6–12 mm wide; ligule 7–15 mm ..	**38 laevigata**
67b	Leaves 3–6 mm wide; ligule 1–2 mm ..	**39 binervis**
68a (51)	Sheaths and usually lower surface of lvs pubescent	**48 pallescens**
68b	Plant glabrous	69
69a (68)	Lowest bract not sheathing	70
69b	Lowest bract with a cylindrical sheath at least 1 mm	75
70a (69)	Utricles papillose	**34 flacca**
70b	Utricles smooth	71
71a (70)	Utricles with a notched beak c. 0.5 mm ..	**23 acutiformis**
71b	Utricles with a bifid beak at least 0.75 mm	72
72a (71)	♀ glumes exceeding utricles	**24 riparia**
72b	♀ glumes shorter than utricles	73
73a (72)	Utricles sterile; alpine flushes on mainly inorganic soils	**29 × grahamii**
73b	Utricles fertile; mires and reedswamps, if on mountains then in deep peat flushes ..	74
74a (73)	Utricles 4–6 mm patent, ± abruptly contracted into beak; leaves usually glaucous	**26 rostrata**
74b	Utricles 6–8 mm, erecto-patent, gradually narrowed into beak; leaves usually yellowish-green	**27 vesicaria**
75a (69)	Plant not or only laxly caespitose, with long creeping rhizomes	76
75b	Plant ± densely caespitose, without or with very short creeping rhizomes	80
76a (75)	Utricles papillose	**34 flacca**
76b	Utricles smooth	**77**
77a (76)	Utricles 7–9 mm	**37 depauperata**
77b	Utricles less than 7 mm	78
78a (77)	Utricles with a prominent, scabrid, bifid beak	**43 hostiana**
78b	Utricles without or with a smooth, entire or weakly notched beak	79
79a (78)	Sheath of lowest bract inflated, loose; leaves yellowish to dark green; utricles gradually narrowed into a beak 0.5–1 mm	**36 vaginata**

79b	Sheath of lowest bract not inflated, tight; leaves glaucous; utricle abruptly contracted into a very short beak c. 0.3 mm ..	**35 panicea**
80a (75)	Flowering stems lateral, leafless ..	**50 ornithopoda**
80b	Flowering stems terminal in middle of leaf-rosettes, usually leafy at base	81
81a (80)	Utricles 7–9 mm	**37 depauperata**
81b	Utricles less than 7 mm	82
82a (81)	At least the lower and middle utricles in each spike patent or deflexed :	83
82b	All utricles erecto-patent or appressed ..	88
83a (82)	♀ glumes with prominent, wide silvery scarious margin; apex of sheath of stem lvs with a projection opposite the ligule ..	**43 hostiana**
83b	♀ glumes without prominent, scarious margin; apex of sheath of stem lvs without a projection	84
84a (83)	♀ spikes oblong to cylindrical , ±distant ..	**41 punctata**
84b	♀ spikes ovoid, on some stems at least the 2 upper approximate	85
85a (84)	Beak of utricles curved or deflexed, usually at least ½ as long as the usually ± curved body	86
85b	Beak of utricles neither deflexed nor curved, usually less than ½ as long as the ± straight body	87
86a (85)	Utricles 6–6.5 mm; ♂ spike usually subsessile; lvs up to 7 mm wide, at least ⅔ as long as stems	**44 flava**
86b	Utricles 3–5 mm; ♂ spike usually pedunculate; lvs not more than 4 mm wide, less than ⅔ as long as stems..	**45 lepidocarpa**
87a (85)	Utricles 3–4 mm, gradually narrowed into a beak c. 1 mm long; ♂ spike on a peduncle 3–25 mm	**46 demissa**
87b	Utricles 1.75–3.5 mm, abruptly contracted into a beak 0.5–1 mm or gradually narrowed into a minute beak; ♂ spike usually sessile	**47 serotina**
88a (82)	Apex of sheath of stem-lvs truncate opposite the ligule; ♀ spikes 2–3 mm wide, lax	**33 strigosa**

88b	Apex of sheath of stem-lvs with a projection opposite the ligule; ♀ spikes at least 4 mm wide, dense	89
89a (88)	Lvs channelled or with inrolled margins; lowest bract equalling or exceeding infl. ..	**42 extensa**
89b	Lvs flat; lowest bract shorter than infl. ..	90
90a (89)	At least some lvs on the plant more than 6 mm wide; ligule more than 5 mm ..	**38 laevigata**
90b	Lvs not more than 6 mm wide; ligule less than 5 mm	91
91a (90)	♀ glumes with prominent, wide, silvery scarious margin; leaf-blade abruptly contracted below the linear, veinless apex	**43 hostiana**
91b	♀ glumes without prominent scarious margin; leaf-blade gradually narrowed to the apex	92
92a (91)	Basal sheaths orange-brown; ♀ glumes dark reddish- or purplish-brown	**39 binervis**
92b	Basal sheaths pale to dark brown, not orange; ♀ glumes pale brown or pale reddish-brown	**40 distans**

KEY TO NON-FLOWERING SPECIMENS OF CAREX AND OTHER SIMILAR CYPERACEAE

1a	Plants without obvious long rhizomes, forming tussocks raised on peaty pedestals 10 cm or more high	2
1b	Plants creeping or, if tufted, then tuft not raised above 10 cm	8
2a (1)	Lvs setaceous and trigonous or short and subulate at base of stem	3
2b	Lvs not as above	4
3a (2)	Lvs 30–50 cm, trigonous, the sheaths pink-brown .. .. .. **Eriophorum vaginatum**	
3b	Lvs 5–15 mm at base of stem, the sheaths yellow-brown (young stems appear similar to lvs of above) .. .. .. **Scirpus cespitosus**	
4a (1)	Tussocks with rhizomes with at least 10 internodes, entwined throughout the crown **68 C. nigra**	
4b	Tussocks with no such long rhizomes ..	5
5a (4)	Lvs glaucous or blue-green, ±dull on both sides, plicate; lf-sheaths fibrillose on splitting .. .. **67 C. elata**	
5b	Lvs not glaucous, ± shiny beneath, channelled or flat; lf-sheaths not fibrillose on splitting	6
6a (5)	Lvs 3–7 mm wide, deeply channelled, dark green; basal sheaths and scales not becoming fibrous .. .. .. **1 C. paniculata**	
6b	Lvs 1–3 mm wide, ±flat or keeled, bright or yellow-green; basal sheaths soon decaying leaving black bristly fibres	7
7a (6)	Lvs keeled; largest roots as thick as base of stem .. .. **2 C. appropinquata**	
7b	Lvs ± flat or slightly channelled; roots much thinner than base of stem .. .. **14 C. remota**	
8a (2)	Plants tufted, lacking pioneering rhizomes	9

8b	Plants creeping or if tufted then with pioneering rhizomes connecting tufts ..	57
9a (8)	Lvs hairy, at least on one surface or around the ligule and sheath apex	10
9b	Lvs glabrous	12
10a (9)	Lf-sheaths pubescent	**48 C. pallescens**
10b	Lf-sheaths glabrous	11
11a (10)	Lfy shoots overwintering; infl. stems arising from axils of overwintering lvs; lvs yellow-green; lf-sheaths with bright red blotches, ± fibrous on decay; rhizomes thin	**49 C. digitata**
11b	Lfy shoots not overwintering: infl. stems terminal from dormant shoots; lvs blue- to pale-green; lf-sheaths deep red-brown, very fibrous; rhizomes thick ..	**55 C. montana**
12a (9)	Lf-blades 5–15 mm long, subulate, at base of evergreen stems; no sterile shoots obvious (young stems not to be confused with setaceous lvs)	**Scirpus cespitosus**
12b	Lf-blades larger; if as above then sterile shoots present	13
13a (12)	Lvs setaceous, usually 1 mm wide or less ..	14
13b	Lvs not setaceous, at least some wider than 1 mm..	16
14a (13)	Basal sheaths red-brown and shiny at first, becoming black and dull; lvs subterete	**Schoenus nigricans**
14b	Basal sheaths pale pinkish to deep brown, never black	15
15a (14)	Rhizomes shortly creeping; tuft ± open; scales and sheaths deep brown; lvs V-shaped in TS	**73 C. pulicaris**
15b	Rhizomes densely tufted; scales and sheaths pink-brown; lvs trigonous	
		Eriophorum vaginatum
16a (13)	Lvs greyish or pale green, usually dull, or glaucous at least on one surface or when young	17
16b	Lvs mid- to yellow-green, not glaucous above or beneath	27
17a (16)	At least some inner sheaths with a protruding tongue at apex of inner face	**40 C. distans**

17b	Inner sheath straight to concave at apex of inner face	18
18a (17)	Ligule tubular	**41 C. punctata**
18b	Ligule not tubular	19
19a (18)	Basal sheaths and/or rhizome-scales red-brown	20
19b	Basal sheaths and scales with no trace of red	22
20a (19)	Mountain plants of wet rocky ledges above 600 m (2000 ft); at least a few sheaths fibrillose	**61 C. atrata**
20b	Estuarine or lowland plants below 600 m (2000 ft); sheaths not fibrillose	21
21a (20)	Lvs 2–3 mm wide, often inrolled, grey-green above; a coastal plant ..	**42 C. extensa**
21b	Lvs 15–20 mm wide, not inrolled, yellow-green above; inland plant of woods and scrubs	**30 C. pendula**
22a (19)	Sheaths distinctly fibrillose on splitting; shoots in a dense tussock; basal scales robust, shiny yellow-brown	**67 C. elata**
22b	Sheaths not fibrillose; shoots and basal scales not as above	23
23a (22)	Lvs plicate, soft; rhizome-scales and lf-sheaths pink brown	24
23b	Lvs keeled or channelled, ± stiff; basal sheaths and scales whitish or blackish brown	25
24a (23)	Ligule 2–3 mm, acute; plants of acid bogs and base-poor mires (widespread) ..	**20 C. curta**
24b	Ligule c. 1 mm, rounded; plants of high altitude flushes and ledges (Scotland only)	**19 C. lachenalii**
25a (23)	Inner face of lf-sheath herbaceous except for V-shaped, brown, membranous portion at apex (acid bogs)	**Rhynchospora alba**
25b	Inner face of lf-sheath narrow but hyaline and membranous to base (coastal and estuarine)	26
26a (25)	Ligule tubular; basal sheaths brownish, not fibrous on decay	**47 C. serotina**
26b	Ligule not tubular; basal sheaths blackish brown, fibrous on decay	**42 C. extensa**

27a (16)	Rhizome-scales or lower lf-sheaths red or red-brown	28
27b	Rhizome scales and lower lf-sheaths with no signs of red colouring (old lvs or lf-sheaths of *C. spicata* may have some isolated blotches of a purplish wine-red colour)	38
28a (27)	Plants of shady damp woodlands; inner face of lf-sheath with minute flecks of red-brown pigment; lvs usually 8–15 mm wide (depauperate forms of *C. laevigata* may be as narrow as 4 mm)	29
28b	Plants of more open habitats; lf-sheath without such flecks; lvs less than 7 mm wide	30
29a (28)	Ligule of inner lvs not more than 8 mm, $\pm$ obtuse; lvs thin, flaccid	**33 C. strigosa**
29b	Ligule of inner lvs usually more than 8 mm, acute; lvs tough, stiff and papery ..	**38 C. laevigata**
30a (28)	Apex of inner face of lf-sheath (especially of inner lvs around stems) with protruding tongue	31
30b	Apex of inner face of lf-sheaths without such a protruding tongue	32
31a (30)	Lvs with patches of wine-red when ageing and turning brown, persisting as an orange-brown litter when dead; lf-sheaths not fibrous	**39 C. binervis**
31b	Lvs turning brown then grey, without transitional red blotches; lf-sheaths fibrous	**40 C. distans**
32a (30)	Lf-margins distinctly rough at base; lvs retaining colour on drying; plants of chalk and limestone downland or scrub (England and Wales only)	33
32b	Lvs $\pm$ smooth at base (in mountain plants sometimes slightly rough) tending to dry bluish green; plants of acid soils or on mountains	36
33a (32)	Terminal shoot always lfy, yellow-green; lateral shoots with flowering stems and few lvs; rhizome not thickly clothed in fibrous scales; roots c. 0.5 mm thick	34

33b	Terminal shoots eventually bearing infl.; lateral shoots lfy, mid-to dark-green, the basal sheaths becoming fibrous on decay giving thick appearance to rhizome; at least some roots 1 mm thick	35
34a (33)	Lf-sheaths usually fibrous; basal lfless sheaths of lateral shoots deep wine-red	**49 C. digitata**
34b	Lf-sheaths not or only slightly fibrous; basal lfless sheaths of lateral shoots green, rarely red-tinged	**50 C. ornithopoda**
35a (34)	Lvs soft, flat, dying at end of season; lower sheaths russet red-brown; roots grey-brown	**55 C. montana**
35b	Lvs stiff, becoming channelled and recurved, overwintering; lower sheaths patchily orange or red-brown; roots dark brown	**51 C. humilis**
36a (32)	Lower lf-sheaths not fibrous, red-brown (rare, high Scottish mountains) ..	**63 C. norvegica**
36b	Lower lf-sheaths markedly fibrous, fibres pale brown	37
37a (36)	Lvs c. 2 mm wide, often up to 20 cm long, smelling of turpentine	**56 C. pilulifera**
37b	At least some lvs 0.5–1 mm wide, rarely over 10 cm long, not smelling of turpentine	**32 C. capillaris**
38a (27)	Lvs more than 7 mm wide; plants of dyke-sides and wet places	39
38b	Lvs rarely more than 5 mm wide, if so then soft and flaccid; plants of various habitats	41
39a (38)	Lower lf-sheaths fibrillose on splitting, not fibrous; or if leaving vascular strands then such fibres whitish; lvs yellow-green	**25 C. pseudocyperus**
39b	Lower lf-sheaths not fibrillose on splitting, forming on decay blackish-brown fibres; lvs dark- or mid-green	40
40a (39)	Plants usually standing in water; hyaline inner face of lf-sheath brown-flecked, with transverse wrinkles; ligule truncate, broader than the leaf	**4 C. vulpina**
40b	Plants rarely standing in water; hyaline inner face of lf-sheath not brown-flecked, without transverse wrinkles; ligule acute, narrower than the leaf	**5 C. otrubae**

41a (38)	Internodes of lfy shoots contracted at base; lf-sheaths not forming a false stem ..	42
41b	At least some internodes of lfy shoots extended (nodes can be felt with finger and thumb); lf-sheaths forming a false stem ..	49
42a (41)	Plants of shaded well-drained situations; inner face of lf-sheath pale (orange-) brown; lvs flat, soft, usually tapering below	**31 C. sylvatica**
42b	Plants of wet places; inner face of lf-sheath hyaline, or if brown, then a rare plant of bogs; lvs keeled, channelled or flat, widest at base	43
43a (42)	Lvs 1–1.5 mm wide, ± erect; shoots often less than 5	44
43b	Lvs 2–7 mm wide, often arcuate; shoots usually more than 5	45
44a (43)	Lf-sheaths a rich orange-brown; shoot-buds not bulbous; inner face of sheath wide, membranous throughout	**Kobresia simpliciuscula**
44b	Lf-sheaths pale or whitish brown ensheathing bulbous shoot-buds; inner face of sheath herbaceous with small V-shaped, membranous apex ..	**Rhynchospora alba**
45a (43)	Lvs 4–7 mm wide; ligule c. 5 mm, acute ..	**44 C. flava**
45b	Lvs 2–4 mm wide; ligule not more than 3 mm, obtuse or rounded	46
46a (45)	Lvs flat or channelled but midrib not prominent beneath; ligule tubular	**47 C. serotina**
46b	Lvs deeply keeled or midrib at least prominent beneath; ligule not tubular ..	47
47a (46)	Lvs 2–3 mm wide, ± thick; septa in lf-sheaths obscure; ligule incurved and stiff; lowest lf-sheaths dark brown	**16 C. echinata**
47b	Lvs (2.5–) 3–5 mm wide; septa in lf-sheaths conspicuous; ligule neither incurved nor stiff; lowest lf-sheaths pale..	48
48a (47)	Lvs bright yellow-green, rather rough on margins and back of midrib; plants of 2–3 shoots, laxly tufted	**45 C. lepidocarpa**
48b	Lvs dark yellow-green, almost smooth; plants usually of more than 3 shoots, ± densely tufted	**46 C. demissa**

49a (41)	Rhizome-scales and lf-sheaths not fibrous, shiny, pale or pink-brown	**18 C. elongata**
49b	Rhizome-scales and lf-sheaths fibrous, not shiny or pink-brown	50
50a (49)	Plants tending to form a definite, even if small, raised tussock; lvs 1.5–2.5 mm wide, channelled, distal part drooping	**14 C. remota**
50b	Plants tufted but close to ground; lvs keeled or flat, usually wider than 2.5 mm, not drooping	51
51a (50)	Lfy shoots ± decumbent, forming a spreading tuft open in the centre	**15 C. ovalis**
51b	Lfy shoots erect, forming a close, ±dense tuft	52
52a (51)	Lvs abruptly contracted on insertion into sheaths, ± auriculate; ligule tubular ..	**5 C. otrubae**
52b	Lvs not contracted as above; ligule not tubular	53
53a (52)	Ligule acute, 4–10 mm long; roots grey-purple	**6 C. spicata**
53b	Ligule ovate, rounded or truncate, less than 4 mm long; roots yellowish grey-brown	54
54a (53)	Ligule rounded or truncate, as long as broad, or shorter	55
54b	Ligule nearly ovate, slightly longer than broad..	56
55a (54)	Sheath splitting easily from the stem; ligule whitish, rounded, as long as wide	**8a C. divulsa** subsp. **divulsa**
55b	Sheath not splitting easily from the stem; ligule yellowish, usually wider than long	**8b C. divulsa** subsp. **leersii**
56a (54)	Plants growing in limestone scree (very rare in Britain) ..	**7a C. muricata** subsp. **muricata**
56b	Plants growing on acid sands or gravels or on stone banks (frequent in Britain)	**7b C. muricata** subsp. **lamprocarpa**

72b	Rhizomes not as above; at least outer lf-sheaths reddish	73
73a (72)	Ligule 1–2 mm, obtuse; red coloration on outer lf-sheaths only (a local plant of grassy places, S and W England) ..	**53 C. tomentosa**
73b	Ligule c. 3 mm, acute; red coloration on both outer and inner lf-sheaths (a rare plant of Scottish mesotrophic mires)	**62 C. buxbaumii**
74a (66)	Lvs bluish or grey-green or glaucous at least on one surface	75
74b	Lvs not bluish, grey-green or glaucous ..	82
75a (74)	Lvs thick, tough, with sharp teeth on margins; rhizome-scales persistent, red-brown, with hyaline margins at base	**Cladium mariscus**
75b	Lvs not as above, if serrate then not thick and tough; rhizome-scales without hyaline margins at base	76
76a (75)	Lvs stiff, usually arcuate, glaucous; rhizome-scales shiny, red- or purple-brown, persisting for some time	**66 C. bigelowii**
76b	Lvs not as above, or if so then dark green on upper surface; rhizome-scales not as above, soon decaying	77
77a (76)	Lvs 6–15 mm wide, tips ±erect, abruptly tapered at apex	**24 C. riparia**
77b	Lvs 2–5 mm wide, distal part ±drooping gradually tapered and often acicular at apex	78
78a (77)	Lvs distinctly paler and glaucous beneath, dark or mid-green above	**34 C. flacca**
78b	Lvs similar on both surfaces or darker beneath	79
79a (78)	Lvs acicular at tip; lf-sheaths thick and spongy, having the appearance of corrugated cardboard; ligule rounded	**26 C. rostrata**
79b	Lvs not acicular at tip· lf-sheaths thin, not wrinkled; ligule acute	80
80a (79)	Lvs up to 25 cm, 1.5–3 mm wide, with c. 9 veins; lf-sheaths becoming fibrous	**59 C. rariflora**
80b	Lvs usually over 25 cm and at least 3 mm wide (if smaller then shiny beneath); lf-sheaths persistent	81

81a (80)	Lvs dull, plicate; lf-sheaths rarely reddish tinged, then only at base; split hyaline inner face persistent as a brown strip; stomata on lower surface of lf only ..	**69 C. acuta**
81b	Lvs flat or channelled, shiny beneath; lf-sheaths orange-red above substrate- or water-level, the inner face soon shrivelling when split; stomata on upper surface of lf only	**65 C. aquatilis**
82a (74)	Lvs thin and stiff (papery) tending to droop at tip; at least the upper lf-sheaths with protruding tongue at apex of inner face ..	83
82b	Lvs thick or if thin then not stiff and drooping at tip; inner face of lf-sheath without protruding tongue	84
83a (82)	Lvs 2–6 mm wide, dark green; ligule c. 1.5 mm (a plant of open heaths and mountain pastures)	**39 C. binervis**
83b	Lvs 5–10 mm wide, mid-green; ligule 7–15 mm (a plant of damp woodland)	**38 C. laevigata**
84a (82)	Plants of mountain flushes or wet ledges or of moorland or *Sphagnum* bogs	85
84b	Plants not in the above habitats	91
85a (84)	Lvs dark green at least above, thick and shiny, with a long trigonous point; lf-sheaths ±uniformly orange-, purple- or red-brown, usually spongy	86
85b	Lvs not as above, if with trigonous tips then abruptly tapered; lf-sheaths streaked with red- or purple-brown, not spongy ..	88
86a (85)	Lvs with wine-red splotches on dying; trigonous tip more than 50 mm long **Eriophorum angustifolium**	
86b	Lvs often yellowish to straw-coloured on dying, never blotched with red; trigonous tip not more than 50 mm long	87
87a (86)	Lvs c. 2 mm wide ±channelled; lf-sheaths orange-red	**28 C. saxatilis**
87b	Lvs 3–4 mm wide, flat; lf-sheaths purple-red	**29 C. × grahamii**
88a (85)	Roots white, not felty; lvs abruptly tapered to trigonous blunt point	**36 C. vaginata**

88b	Roots covered with a conspicuous yellow felt; lvs gradually narrowed to a fine blunt point	89
89a (88)	Ligule 3–4 mm obtuse, not tubular; lower internodes of sterile shoots often elongating and becoming 'stoloniferous'; inner sheaths blotchy red; rhizome-scales usually brown	**58 C. limosa**
89b	Ligule 5–8 mm, acute, tubular; sterile shoots not 'stoloniferous'; inner sheaths pale orange-brown; rhizome-scales reddish	90
90a (89)	Free portion of ligule c. 3 mm; lf-sheaths more readily fibrous; lvs with c. 9 veins	**59 C. rariflora**
90b	Free portion of ligule less than 1 mm; lf-sheaths persistent; lvs with c. 17 veins	**60 C. magellanica**
91a (84)	Inner face of lf-sheath not hyaline; lf-sheaths thick, fibrous, the septa persistent on decay	**Cyperus longus**
91b	Inner face of lf-sheath hyaline; lf-sheaths not as above	92
92a (91)	Lvs 4–7 mm wide; plants of wet riverine situations in Scotland	**64 C. recta**
92b	Lvs 1–4 mm wide; plants of usually dry calcareous situations in England and Wales	93
93a (92)	Lvs up to 2 mm wide; lf-sheaths and rhizome-scales dull; plants of open grassland	**54 C. ericetorum**
93b	Lvs 2–4 mm wide; lf-sheaths and rhizome-scales shiny and red-tinged; woodland plants	**37 C. depauperata**
94a (63)	Basal node of shoot swollen and bulb-like	**Scirpus maritimus**
94b	Basal nodes not swollen	95
95a (94)	Lower lf-sheaths prominently reticulate-veined, stout	96
95b	Lower lf-sheaths not reticulate-veined, slender	97
96a (95)	Rhizomes shortly creeping, forming loose tussocks at ±regular intervals	**3 C. diandra**
96b	Rhizomes far-creeping, often branched ..	**12 C. divisa**

97a (95)	Lf-tips tapered into a fine acicular point; sheaths very thick and spongy, having appearance of corrugated cardboard; rhizome white	**26 C. rostrata**
97b	Plants without the above combination of characters	98
98a (97)	Riparian or marsh plants usually in large stands; lvs at least 5 mm wide	99
98b	Plants in various habitats; lvs 4 mm wide or less	101
99a (98)	Lvs abruptly tapered, ±erect at tip, glaucous on both sides	**24 C. riparia**
99b	Lvs long-tapered with drooping tips ..	100
100a (99)	Lower lf-sheaths fibrillose; inner face soon breaking and shrivelling; old lvs dark-, almost reddish-green above ..	**23 C. acutiformis**
100b	Lower lf-sheaths not fibrillose; inner face hyaline and if splitting remaining as a brown papery strip; all lvs glaucous ..	**69 C. acuta**
101a (98)	Plants of sandy or estuarine shores ..	102
101b	Plants of non-saline habitats	106
102a (101)	Lvs not glaucous or blue green; plant usually within tide or spray zone	103
102b	Lvs glaucous or blue-green at least beneath; if near sea then in runnels of fresh water	104
103a (102)	Lvs channelled; rhizome tough, woody; shoots at least 8 cm apart; roots wiry, with numerous secondary branches ..	**13 C. maritima**
103b	Lvs inrolled; rhizome fleshy; shoots usually less than 5 cm apart; roots thick, aerenchymatous, with few secondary branches	**Blysmus rufus**
104a (102)	Lvs dark green above, pale bluish beneath, with 13–30 closely packed veins; septa in lf-sheaths conspicuous	**34 C. flacca**
104b	Lvs equally glaucous on both sides, with 7–15 well-spaced veins; septa in lf-sheaths obscure	105
105a (104)	Inner face of lf-sheath brown, papery and persistent; lf-margin rolling outwards on drying, stomata confined to the lower surface	**69 C. acuta**

105b	Inner face of lf-sheath transparent, soon decaying; lf-margin rolling inwards on drying, stomata ± confined to the upper surface	**68 C. nigra**
106a (101)	Lf-sheaths or rhizome-scales deep or chestnut-brown; plants of well drained base-rich sites	107
106b	Lf-sheaths or rhizome-scales blackish or grey-brown to whitish; plants of marshes, mires and bogs	108
107a (106)	Lf-sheaths distinctly fibrous; inner sheaths whitish; lvs not flexuous	**52 C. caryophyllea**
107b	Lf-sheaths not fibrous; inner sheaths often orange; lvs flexuous	**72 C. rupestris**
108a (106)	Lvs with a distinct thick, trigonous, subulate tip; midrib channel ending abruptly some distance below apex	109
108b	Lf-tip not trigonous, or if so then very slender and acicular; midrib channel running gradually out to apex	112
109a (108)	Basal lf-sheaths and rhizome-scales brown, soon becoming fibrous; inner face of lf-sheath herbaceous, only the apex membranous	**Eriophorum latifolium**
109b	Basal lf-sheaths and rhizome-scales not as above; inner face membranous to base ..	110
110a (109)	Rhizome fleshy, scales blackish, ± persistent; subulate tip of lf 1.5–3 cm long	**Blysmus compressus**
110b	Rhizome not fleshy; scales grey-brown, soon decaying; subulate tip less than 1 cm	111
111a (110)	Plant distinctly glaucous; apex of inner face of all sheaths either straight or concave	**35 C. panicea**
111b	Plant light to yellowish green, rarely glaucous; apex of inner face of at least the middle lf-sheaths with a protruding tongue	**43 C. hostiana**
112a (108)	Lvs glaucous when young at least beneath, gradually tapered, apex not acicular ..	113
112b	Lvs yellowish green when young; apex usually blunt or abruptly tapered, or if gradually tapered then acicular	114
113a (112)	Lvs glaucous on both surfaces; veins 7–15, well-spaced; septa in lf-sheath obscure ..	**68 C. nigra**

113b	Lvs pale bluish beneath, darker green above; veins 13–30, close-packed; septa in lf-sheath conspicuous	**34 C. flacca**
114a (112)	Plant of mesotrophic mires (Sutherland and Inverness only); sterile shoots arising singly from lower decumbent nodes of flowering stem	**11 C. chordorrhiza**
114b	Plant not as above	115
115a (114)	Base of shoot terete, 7–12 mm thick; lf-sheaths brown, fibrous, septa persistent on decay	**Scirpus sylvaticus**
115b	Base of shoot trigonous, 2–4 mm thick; lf-sheaths white or pink-brown, rarely fibrous; septa not persistent	116
116a (115)	Lf-apex drawn out into an acicular point; shoots decumbent at base or if erect then lvs 25 cm or more	**19 C. lachenalii**
116b	Lf-apex not drawn out into acicular point; shoots erect, 20 cm or less	117
117a (116)	Rhizomes yellow, far-creeping; lvs of sterile shoots 2–2.5 mm wide, apex with a trigonous tip	**36 C. vaginata**
117b	Rhizomes whitish, short, with few shoots; lvs of sterile shoots often less than 1 mm wide, apex blunt but not trigonous	**57 C. atrofusca**
118a (57)	Lfy shoots overwintering, borne ± constantly at every 4th node; inner face of lf-sheath hyaline, thin and easily split	**9 C. arenaria**
118b	Lfy shoots not overwintering, not regularly spaced; inner face of lf-sheath herbaceous, not easily splitting	**10 C. disticha**

DESCRIPTIONS AND FIGURES

For obvious reasons the descriptions of the species have been kept as short as possible, and certain conditions have been accepted as standard and their absence or alternative only is mentioned. The exception to this is, in closely related species, where a diagnostic character has been compared throughout. Unless otherwise stated the following can be taken as the general state: *scales* and *leaf-sheaths* are dull; *stems* are smooth, hollow and leafy at the base; *Carex* plants are glabrous and are hemicryptophytes dying down at the onset of winter; *roughness* in stems and leaves means that they are serrulate on angles or margins and veins; *male spikes* are fusiform; *utricles* are not shiny. The length of the stem is that of the fruiting stem including the inflorescence; the length of the ligule (which can best be seen on the flowering stem) is measured from the level at which the leaf-sheath joins the leaf-lamina to the apex of the ligule (and is not just the length of the free part of the ligule); utricle length includes the beak. The figures given after "Fr." at the end of the description indicate the month(s) of the fruiting period. The figures, with few exceptions, which are otherwise stated on the plate, are drawn to the following scales:

A: whole plant, $\times \frac{1}{2}$

B: transverse section of stem, $\times$ 15

C: transverse section of leaf, $\times$ 10

D: ligule and sheath apex, $\times$ 3

E: part inflorescence, $\times$ 1

F: male floret, $\times$ 6

G: female floret, $\times$ 6

H: utricle, $\times$ 6

J: dissected utricle showing nut, $\times$ 6

1. Carex paniculata L.

Greater Tussock Sedge   Map 1

Rhizomes very short forming dense tussocks up to 1.5m high and 1 m diameter; roots thick, dark brown, felty; scales dark brown, ribbed, persistent. *Stems* 60–150 cm, rough, trigonous, spreading. *Lvs* 20–120 cm × 5–7 mm, stiff, channelled or involute, tapering abruptly to a short trigonous tip, dark green, shiny beneath, over-wintering; margins serrulate; sheaths of sterile shoots forming a false stem, lowermost sheaths brown with shorter blades, inner face hyaline, apex brown, concave; ligule 2–5 mm, rounded. *Infl.* a compact panicle, 5–15 cm. *Spikes* numerous, mostly compound, the lower clusters pedunculate; spikelets 5–8 mm, ♂ at top, ♀ below, some lower spikelets all ♀; bracts setaceous or glumaceous with broad hyaline margin and excurrent midrib. ♂ *glumes* 3–4 mm, ovate-lanceolate, orange-brown, hyaline, with pale midrib; apex acute. ♀ *glumes* 3–4 mm, ovate-triangular, appressed to utricle, orange-brown, with wide hyaline margin; apex acute or acuminate. *Utricles* 3–4 mm, ovoid, ± trigonous, ventricose at base, ribbed, green to dark brown; beak 1–1.5 mm, winged, serrulate, deeply split on adaxial side; stigmas 2; nut ovoid, biconvex.   Fr. 6–8.

C. paniculata is a species of peaty, medium base-rich soils where water levels are at least seasonally high; in fens and beside slow-flowing streams. An important component of the early hydrosere facilitating establishment of tree spp.; it can withstand a certain amount of the resulting shade. Common throughout Ireland and Britain but becoming less frequent in Scotland due to lack of suitable lowland habitats.

This distinct sedge can be confused only with No. 2. Recorded from Britain are sterile hybrids with that species (= *C.* × *solstitialis* Figert), with *C. remota* (= *C.* × *boenninghausiana* Weihe) and *C. otrubae*; hybrids with *C. curta*, *C. diandra*, *C. echinata*, *C. elongata* and *C. vulpina* are reported from Europe.

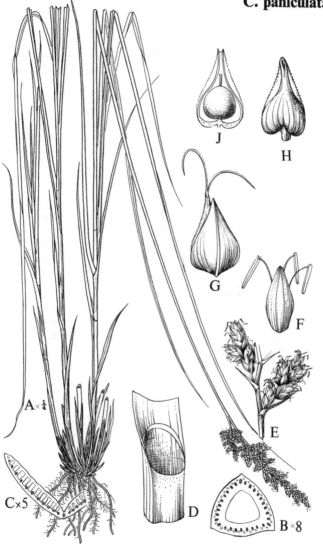

A × ¼

C × 5

D

B × 8

E

F

G

H

J

2. Carex appropinquata Schumacher

Fibrous Tussock Sedge Map 2

Rhizomes very short forming dense tussocks up to 1 m high and 80 cm diameter; roots thick, dark brown, felted; scales black, soon breaking up into wiry fibres. *Stems* 40–80 cm, rough, trigonous. *Lvs* 20–80 cm × 1–2 mm, stiff, keeled, yellow-green; ligule 2–3 mm, rounded. *Infl.* a compact panicle 4–8 cm; spikelets 4–7 mm, ♂ at top, ♀ below, some lower spikelets all ♀ and ± pedunculate. ♂ *glumes* 3–4 mm, ovate, red-brown, hyaline, apex acute. ♀ *glumes* 3–4 mm, ovate, acuminate, red-brown-hyaline. *Utricles* 3–4 mm, ovoid to subglobose, abruptly narrowed into the beak; beak 1–1.5 mm, rough, not winged, notched; stigmas 2; nut ovoid, biconvex.   Fr. 5–7.

A plant of similar habitats to *C. paniculata* with which it sometimes grows, but perhaps requiring less water movement and occasionally in more acid fen. Frequent in East Anglia, and formerly (now extinct) on the Middlesex-Herts-Bucks borders; a single station in Pembroke; very local in SE and Mid-West Yorks, and in Roxburgh and Selkirk; widespread but local in central Ireland (Westmeath, Offaly, Clare, formerly Carlow). Possibly sometimes overlooked or confused with *C. diandra* (see below) or with depauperate specimens of *C. paniculata* from which the black, fibrous sheaths, the red colouring of the glumes (more pronounced on drying), and the unwinged utricle should sufficiently distinguish it. *C. appropinquata* hybridises in England with *C. paniculata* (*C.* × *solstitialis* Figert); hybrids with *C. curta, C. diandra,* and *C. remota* are recorded in Europe.

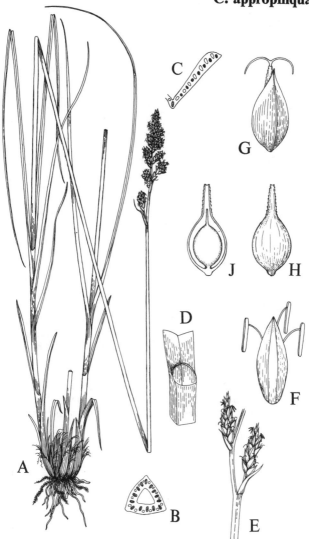

3. Carex diandra Schrank

Lesser Tussock Sedge   Map 3

Rhizomes shortly creeping; shoots clustered, forming a loose tussock or appearing at ± regular intervals; roots grey- or orange-brown; scales black or dark grey-brown, persistent. *Stems* 25–60 cm, slender, ± sharply trigonous, rough on angles. *Lvs* 20–40 cm × 1–2 mm, rough at least above, ± flat or slightly keeled, gradually tapered to a fine ± trigonous point, grey-green, often overwintering; lower sheaths grey- or dark brown, persistent, inner face hyaline, apex ± straight; ligule 1–2 mm, obtuse. *Infl.* a ± compact head, 1–5 cm; bracts glumaceous, lowest rarely setaceous. *Spikes* 6–10, 5–8 mm, sessile, ♂ at top, ♀ below. ♂ *glumes* 3–4 mm, lanceolate-elliptic, pale brown-hyaline; apex acute. ♀ *glumes* c. 3 mm, broadly ovate, pale purple brown, with a short green midrib and broad hyaline margin; apex acute or mucronate. *Utricles* 3–4 mm, broadly ovoid or suborbicular, plano-convex, distinctly 2–5 ribbed in lower half, dark brown, shiny; beak 1.5–2 mm, broad, serrulate, bifid; stigmas 2; nut top-shaped, plano-convex, stalked.   Fr. 6–7.

C. diandra is a species of very wet peaty meadows, alder-sallow carr and more acid swamps often in overgrown ditches or in old peat cuttings. Scattered throughout the British Isles north to Caithness, absent from the SW peninsula and in Ireland mainly in the centre. Predominantly a lowland plant, infrequent in the calcareous districts of Britain and becoming scarcer as sites are drained.

A distinct species rarely confused with others. Depauperate plants in shade and in stagnant water may be tussocky and appear like *C. appropinquata*; that species, however, has very fibrous lower leaf-sheaths. Robust forms have been called *C. ehrhartiana* Hoppe but these do not warrant taxonomic recognition.

C. diandra hybridises with both *C. appropinquata* and *C. paniculata* but these hybrids are not recorded from Britain.

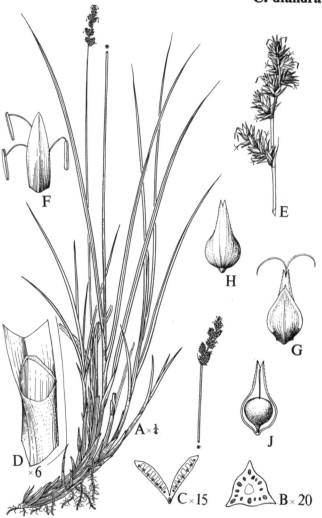

A × ¼

D × 6

F

E

H

G

J

C × 15

B × 20

4. Carex vulpina L.

True Fox Sedge Map 5

Rhizomes short, shoots stout, densely tufted; roots thick, grey-brown; scales dark brown, remaining as black fibres. *Stems* 30–100 cm, smooth below, rough above, sharply trigonous, their faces ± concave and angles winged. *Lvs* up to 80 cm × 4–10 mm, ± erect, keeled, ± abruptly tapering to a flat, sharp point, dark green even when dry, not auriculate at base, inner face of sheath glandular and transversely wrinkled; ligule 2–6 mm, truncate, overlapping the edges of the leaf-blade. *Infl.* a stout, dense panicle; bracts short, with ± prominent dark auricles. *Spikes* numerous, 8–14 mm, compound, upper ♂, lower ♀. ♂ *glumes* 3.5–4 mm, oblanceolate-elliptic, dark or rusty brown with green midrib. ♀ *glumes* 4–5 mm, ovate, dark or rusty brown. *Utricles* 4–5 mm, ovate-elliptic, minutely papillose, readily dropping at maturity; beak 1–1.5 mm, split on back; stigmas 2; nut oblong-obovoid, biconvex.   Fr. 7–8.

In damp places, often standing in water in ditches, usually on heavy clay soils. Mainly in the SE of England, but local; also in Isle of Wight, Berks, Oxford, Gloucestershire, and SE and NE Yorks. Probably overlooked as a result of confusion with *C. otrubae* from which it is most certainly distinguished by the very broad, truncate ligule (in *C. otrubae* it is narrower than, or contained within, the width of the leaf) and the split beak to the papillose utricle. The strongly winged stem, the dark auricles of the bracts, and the glandular inner face of the leaf-sheath may also be diagnostic. Under the microscope, *C. vulpina* has the epidermal cells of the utricle thick-walled and ± square or roundish, while those of *C. otrubae* are thin-walled and oblong.

Possible hybrids with *C. paniculata* and *C. remota* are recorded, but not for the British Isles.

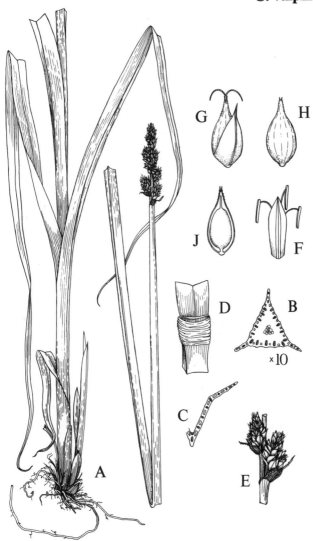

5. Carex otrubae Podp.

False Fox Sedge   Map 6

Rhizomes short; shoots stout, densely tufted; roots thick, grey-brown; scales brown, remaining as black fibres. *Stems* 30–100 cm, smooth below, rough above, trigonous, their faces ± flat, hardly winged on angles. *Lvs* up to 60 cm × 4–10 mm, ± erect, ± auriculate, keeled, ± abruptly tapering to a flat, sharp point, bright green, becoming grey-green when dry and pale orange-brown on decay; margins rough; sheaths white with green veins, becoming brown, soon decaying, inner face hyaline not wrinkled, apex straight; ligule 5–10 mm, ± acute, tubular, contained within the width of the leaf. *Infl.* an elongate panicle, becoming dense when in fruit; lower bracts setaceous, with lf-like base, about as long as infl., upper glumaceous. *Spikes* numerous, 8–14 mm, compound, upper ♂, lower ♀. ♂ *glumes* 3.5–4 mm, oblanceolate-elliptic, pale orange-brown with green midrib; apex acute. ♀ *glumes* 4–5 mm, ovate, pale red- or orange-brown, midrib green; apex acuminate. *Utricles* 5–6 mm, ovate, plano-convex, ribbed, green, dark brown at maturity, not readily dropping; beak 1–1.5 mm, rough at apex, bifid, not split at back; stigmas 2; nut oblong-obovoid, biconvex.

Fr. 7–9.

C. otrubae is a species of heavy soils usually in damp situations, e.g. in roadside ditches and beside dykes. Uncommon on the sandy soils of Norfolk and Suffolk; rarely if ever on peat and consequently markedly absent from the uplands of Scotland, N England, Wales and SW peninsula (from the New Forest westward), becoming in these regions almost exclusively a coastal plant. Scattered throughout Ireland, but commonest near the coast.

Similar to No. 4; narrow leaved plants are difficult to separate from *C. muricata* agg., but the acute tubular ligule and ± auriculate leaves of *C. otrubae* will usually distinguish it. *C. otrubae* hybridises with *C. divulsa*, *C. paniculata*, *C. spicata* (= *C.* × *haussknechtii* Senay) and *C. remota* (= *C.* × *pseudaxillaris* K. Richter).

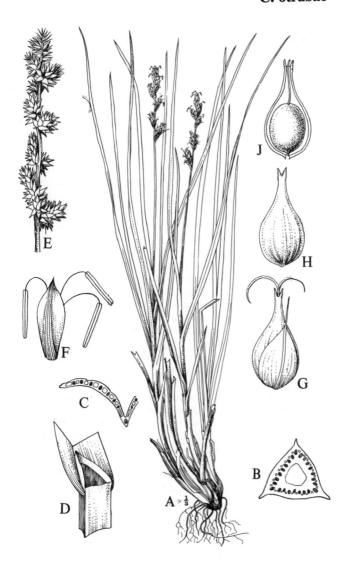

6-8. Carex muricata group

6. C. spicata Hudson
(=*C. contigua* Hoppe; *C. muricata* auct. plur., non L.)

7a. C. muricata L. subsp. **muricata**
(=*C. pairaei* subsp. *borealis* Hyl.)

7b. C. muricata subsp. **lamprocarpa** Čelak
(=*C. pairaei* F. W. Schultz; *C. muricata* subsp. *pairaei*
(F. W. Schultz) Čelak.; *C. bullockiana* Nelmes)

8a. C. divulsa Stokes subsp. **divulsa**

8b. C. divulsa subsp. **leersii** (Kneucker) Walo Koch
(=*C. polyphylla* Kar. & Kir.; *C. leersii* F. W. Schultz, non
Willd.; *C. muricata* subsp. *leersii* Ascherson & Graebner)

The *Carex muricata* L. aggregate (section *Phaestoglochin* Dumort.)
comprises, in Britain, five closely related taxa (see above) whose
precise taxonomic status is debatable. They are most conveniently
considered as falling into three groups: a species standing a little
apart from the rest, and two separate pairs. The most distinct is
C. spicata Hudson, which differs in the structure of utricle and ligule
and in the presence (sometimes only in the roots) of a purplish
pigment wholly absent from the other taxa. The two variants of
C. muricata, treated here as subspecies, show small, but marked and
constant, differences and appear to be ecologically and
geographically vicarious, the one northern and eastern, early-
flowering, and a calcicole, the other western and southern, late-
flowering, and calcifuge. From these three, *C. divulsa* subsp.
divulsa and subsp. *leersii* (*C. polyphylla* Kar. & Kir.) are separated by
the markedly interrupted inflorescence and by the diamond-shape
of the utricles, narrowed both above and below. The extreme form
of *C. divulsa* subsp. *divulsa*, with long, lax inflorescence, appressed
spikes, and small utricles, is strikingly distinct from the extreme
form of subsp. *leersii* which has a shorter, stouter inflorescence and
strongly patent or divaricate utricles of (in Britain) 4 to 4.5 mm.
There are also differences in flowering time, as in the two subspecies
of *C. muricata*. Many intermediates between these extreme forms
can, however, be found. Some of these show a considerable degree
of sterility and may be hybrids between the two, or between one of
them and some other member of the aggregate. Furthermore, *C.
divulsa* subsp. *leersii*, if severely damaged by mowing or other

disturbance, may produce late depauperate flowering stems that are indistinguishable from those of *C. divulsa* subsp. *divulsa*. The two taxa, when considered over their whole geographical range, would appear to represent a cline, or continuous variation, between the extreme form of subsp. *leersii* (*C. polyphylla* of Central Asia, a much more robust plant than any seen in the west, although Karelin and Kirilov's type is far from being an extreme form and can easily be matched by British material) and the subsp. *divulsa* of Western Europe. In these circumstances it seems best to regard these as subspecies and not as species as they have often been referred to in literature.

6.   Carex spicata   Hudson

Spiked Sedge   Map 7

Rhizomes short; shoots densely tufted; roots purple; scales brown or red-brown, with black veins, eventually becoming fibrous. *Stems* 10–85 cm, often stout, acutely trigonous. *Lvs* 7–45 cm × 2–4 mm, keeled, gradually tapering to a flat tip, mid-green; sheaths forming a short false stem, sometimes stained wine-red; apex straight or concave; ligule 4–8 mm, ± acute, with much soft, white, loose tissue. *Infl.* 1–4 cm; bracts glumaceous, with a setaceous point. *Spikes* 3–8, usually contiguous (rarely widely separated), 5–10 mm, sessile, ♂ at top, ♀ below. ♂ *glumes* 3–4 mm, lanceolate, apex acute. ♀ *glumes* 4–4.5 mm lanceolate, acuminate, red-brown with a green midrib. *Utricles* 4–5 mm, with an irregularly bulbous and corky base, and beak 2 mm, narrowly attenuated, scabrid, the whole bronze-green eventually turning dull black; nut rounded-cubical, set well above the base.   Fr. 7–8.

On roadsides and waste ground, and in meadows, more often on the heavier and damper soils than the other members of the group, though also found on the chalk. Frequent throughout southern and especially midland England, becoming less common towards the east coast and rare, possibly only casual, north of the Ribble and Tees, in southern and eastern Scotland, in west Wales, and in Ireland. Absent from Cornwall.

C. spicata is sometimes confused with small specimens of *C. otrubae*, but in that species the usually broader leaves appear (at least in fresh material) pinched where they emerge from the leaf-sheath, many more of the bracts subtending the spikes have long setaceous points, and the utricles are clearly ribbed. From other members of the *C. muricata* group *C. spicata* is distinguished by the long, acute, soft ligule best observed in stem leaves, and the large bulbous-based and narrow-beaked utricles. The acuminate ♀ glumes, which often give the young inflorescence a shaggy appearance, are also distinctive, as is the wine-red pigment which does not appear in any other member of the group and sometimes (but by no means invariably) extends to leaf-sheaths, bracts, and even glumes and which, when present, is definitive.

C. spicata hybridises with *C. echinata* in Europe.

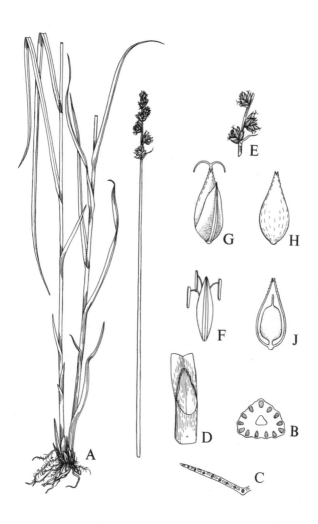

7a. Carex muricata L. subsp. muricata

Prickly Sedge

Map 12

Rhizomes short; shoots densely tufted; roots brown; scales brown with black veins, soon becoming fibrous. *Stems* 10–85 cm, often stout, obtusely trigonous and even polygonal below. *Lvs* 7–45 cm × 2–4 mm, keeled, gradually tapering to a flat tip, mid-green; sheaths forming a short false stem, apex straight or concave; ligule 2–3.5 mm, ovate. *Infl.* 1–4 cm; bracts glumaceous with a setaceous point. *Spikes* 4–8, upper contiguous, lowest separated by up to 10 mm, 5–10 mm, globose, sessile, ♂ at top, ♀ below, or lower entirely ♀. ♂ *glumes* 3–4 mm, elliptic, red-brown, apex acute. ♀ *glumes* 2.5–3.5 mm, markedly shorter than the utricles, dark or red-brown, making at flowering time a strong colour-contrast with the utricles; apex apiculate. *Utricles* 3–4 mm, round in outline with a broad margin, grey-green becoming dark red-brown, shiny; beak abrupt, c. 0.75 mm, minutely toothed; stigmas 2; nut orbicular-compressed. Fr. 6–8.

The commoner subspecies in Scandinavia and eastern Europe but very rare in Britain, on steep, dry, limestone slopes. Confirmed specimens derive from only five British localities, near Woodchester in Gloucestershire, near Wrexham, in Gordale, in Ribblesdale, and at Lauder; but the plant has recently been seen only in the first, second, and fourth of these.

Differs from *C. spicata* in the shape and size of ligule and utricle, from *C. divulsa* in the comparatively compact and dark-coloured inflorescence. From subsp. *lamprocarpa* it is distinguished by the more erect and rigid habit, the early flowering time (May, about a month before subsp. *lamprocarpa*), the marked colour-contrast at this time between the shorter glumes and more rounded utricles (they are concolorous when the fruits mature), and the more globose spikes, the lowest of which is often quite distinctly separated from the next above.

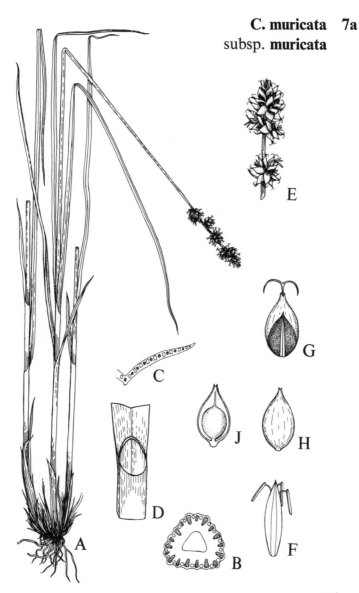

7b. Carex muricata subsp. lamprocarpa Čelak.

Prickly Sedge   Map 8

Rhizomes short; shoots densely tufted; roots brown; scales brown
with black veins, soon becoming fibrous. *Stems* 10–85 cm, often
stout, pentagonal to bluntly trigonous. *Lvs* 7–45 cm × 2–4 mm,
keeled, gradually or sometimes abruptly tapering to a flat tip,
mid-green often bronzing on dying; sheaths forming a short false
stem, apex straight or concave; ligule 2–3.5 mm, ovate. *Infl.* 1–4 mm;
bracts glumaceous with a setaceous point. *Spikes* 3–8, contiguous
or crowded, 5–10 mm, ovoid, sessile, ♂ at top, ♀ below, or lower
entirely ♀. ♂ *glumes* 3–4 mm, lanceolate-elliptic, hyaline or pale
brown, midrib green; apex acute. ♀ *glumes* 3–4.5 mm, ovate, pale
or golden-brown fading to white as the fruits mature; apex acute or
apiculate. *Utricles* 3–4 mm, ovoid with broad base or ovoid-
ellipsoid, yellow-green becoming dark-brown, shiny, tapering
± evenly into a beak; beak c. 0.75 mm, minutely toothed; stigmas
2; nut orbicular-compressed.   Fr. 7–9.

A plant of banks and heaths in open situations, preferring drier,
lighter, and more acid soils and locally abundant in Britain where
these conditions occur but rare on the Midland clays. It is frequent in
southern England and in East Anglia, and common in the south-
west and in west Wales where a favourite habitat is the "west-
country wall". In the north it occurs in quantity in Galloway and on
the Cheviots where it frequently occupies the talus below basalt
cliffs. It becomes rarer again in Scotland north of the Clyde and
Forth, though it reaches Oban in the west and Inverness in the east;
it is uncommon in Ireland.

Robust specimens of *C. muricata* have been regularly confused
with poor specimens of *C. spicata* lacking the distinctive wine-red
coloration; but even when the utricles of *C. muricata* are larger
than usual the absence of an attenuated beak and of any corky
padding at the base, and the short, neatly ovate ligule (to be observed
on stem-leaves), should suffice for a correct determination. The
differences between the two subspecies have been described under
subsp. *muricata*.

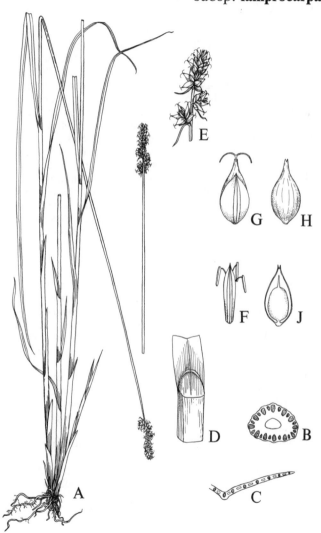

8a. **Carex divulsa** Stokes subsp. **divulsa**

Grey Sedge Map 9

Rhizomes short; shoots in ± dense tufts 10–20 cm across; roots thick, grey-brown; scales dark brown with black veins, soon becoming fibrous. *Stems* 25–75 cm, slender, trigonous, striate. *Lvs* up to 75 cm × 2–3 mm, flat or channelled, gradually tapered to a fine, flat apex, grey-, bronzy or dark green, usually flaccid, often overwintering; sheaths forming false stems in sterile shoots, lower pale brown, ribbed, inner face hyaline, apex straight or ± concave; ligule c. 2 mm, obtuse. *Infl.* 5–18 cm; lower bracts setaceous, 2–3 cm sometimes much longer, upper glumaceous. *Spikes* 5–8, few-fld, upper contiguous, 3–8 cm, all ♀, lower distant from each other by 2 cm or more, sometimes branched, 5–15 mm, ♂ at top and ♀ below. ♂ glumes 3.5–4 mm, lanceolate, ± hyaline; apex acute. ♀ *glumes* 3–3.5 mm, ovate-elliptic, ± hyaline or pale brown with hyaline margins and green midrib; apex acute, often attenuate. *Utricles* 3.5–4 (–4.5) mm, ± erect, ovoid, tapered above and below, faintly nerved at base, shiny, pale yellow or whitish-green turning dull-black; beak c. 1 mm, rough, split; stigmas 2; nut obovoid, biconvex. Fr. 7–9 (–11)

C. divulsa is a plant of hedgerows, wood-borders and waste ground where competition is not too great. It appears relatively indifferent as to soil, and is common throughout southern England from Kent to Cornwall and from the Channel to the Forest of Dean, the Chilterns, and East Anglia, becoming rare in the Midlands but with strong colonies in north-eastern Wales, and Yorkshire east of the Pennines. Found in similar colonies in S and E Ireland.

There is considerable variation in the size of the utricle even on the same spike and the glumes may be tinged with darker brown but on the whole *C. divulsa* is characterised by the colourless glumes and generally pale-coloured, few-fld spikes. From others of the *C. muricata* group (except subsp. *leersii*, see below) *C. divulsa* is distinguished by its elongate inflorescence and the diamond-shaped utricle. In *C. remota* several of the lower bracts are longer than their spikes and often longer than the whole inflorescence.

C. divulsa hybridises with *C. otrubae* in Britain. Hybrids with *C. spicata* and *C. remota* have also been reported but may be no more than aberrant forms of the species. A hybrid with *C. ovalis* is said to occur in Europe.

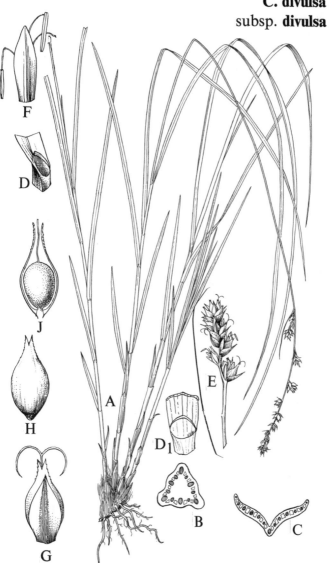

8b. Carex divulsa subsp. leersii (Kneucker) Walo Koch

Leers' Sedge, Many-leaved Sedge   Map 10

Rhizomes short; shoots in $\pm$ dense tufts 10–20 cm across; roots thick, grey-brown; scales dark brown with black veins, soon becoming fibrous. *Stems* 25–90 cm, robust, trigonous. *Lvs* up to 75 cm $\times$ 3–5 mm, flat or channelled, gradually tapered to a fine, flat apex, usually bright yellow-green (darker in shade), stiff; sheaths forming false stems in sterile shoots, lower pale brown, ribbed, inner face hyaline, apex straight or $\pm$ concave; ligule 2 mm or less, obtuse or truncate, often yellow (the blade of the leaf is at this point so closely appressed to the stem that when pulled it parts from it with a snap). *Infl.* 4–8 cm; lower bracts 2–3 cm sometimes longer, setaceous, upper glumaceous. *Spikes* 4–8, upper contiguous, 5–10 mm, all ♀, lower up to 2 cm distant from each other, sometimes branched, 5–15 mm, ♂ at top and ♀ below. ♂ *glumes* 3.5–5 mm, lanceolate, brownish-hyaline; apex acute. ♀ *glumes* 4–4.5 mm, ovate-elliptic, golden-yellow or light-brown; apex acute, often attenuate. *Utricles* 4–4.5 (–4.8) mm, strongly patent, ovoid, tapered above and below, faintly nerved at base, shiny, yellowish soon becoming dark red-brown; beak c. 1 mm, rough, split; stigmas 2; nut obovoid, biconvex.   Fr. 6–8.

In habitats similar to those of subsp. *divulsa*, but strongly calcicole. The most extreme forms are found in Yorkshire, Derby, Rutland, Denbigh and the Cotswolds. Though the plant is scattered over much of the rest of S England and eastern Britain north to Edinburgh, there are also to be found in these parts many populations with characters intermediate between those of subsp. *leersii* and subsp. *divulsa*. In addition damaged plants of subsp. *leersii* may produce weak, late stems resembling those of subsp. *divulsa*.

Characters separating *C. divulsa sensu lato* from other sedges are given under subsp. *divulsa*. From this, subsp. *leersii* in its extreme form differs in the erect, rigid habit, the yellow-green of the foliage, the shorter, stouter inflorescence, and the larger very markedly divaricate utricles which soon turn a shiny red-brown. Subsp. *leersii* flowers in May and has usually shed most of its fruit by the end of August, whereas subsp. *divulsa*, beginning later, has a much longer flowering season from June to October.

9. Carex arenaria L.

*Sand Sedge*Map 11

Rhizomes far-creeping; shoots usually single at about every fourth node; roots pale brown, much branched; scales dark brown, soon becoming fibrous. *Stems* 10–90 cm, rough towards the top, trigonous, often curved, varying in thickness. *Lvs* up to 60 cm × 1.5–3.5 mm, ± flat, rough, rigid, thick, often recurved and keeled or channelled in open habitats, tapering gradually to a fine trigonous point, dark green, shiny, often dark brown on dying; sheaths forming false stems in sterile shoots, lower pale- or grey-brown, persistent, inner face hyaline, becoming brown, membranous, apex straight; ligule 3–5 mm, obtuse, tubular. *Infl.* a dense head up to 8 cm; bracts glumaceous, lower with setaceous points. *Spikes* 5–15, 8–14 mm, terminal spikes ♂, middle ♂ at top, ♀ below, lower entirely ♀. ♂ *glumes* 5–7 mm, lanceolate-elliptic, pale red-brown with hyaline margins; apex acute. ♀ *glumes* 5–6 mm, ovate, pale red-brown, with pale or green midrib and hyaline margins; apex acute or acuminate. *Utricles* 4–5.5 mm, ovate, plano-convex, many-ribbed, broadly winged, serrate in upper half, pale green-brown; beak 1–1.5 mm, winged below, bifid; stigmas 2; nut oblong-ellipsoid.　　Fr. 7–8.

C. arenaria is a dominant plant on fixed dunes and wind-blown sand usually with a low lime content, forming characteristic communities with lichens, *Festuca* spp., *Thymus* and *Calluna-Empetrum nigrum*. Mainly a coastal plant throughout the British Isles but inland in a few localities especially the Brecks and Lincoln heaths.

Only one other British species, *C. disticha*, has the monopodial rhizome system of *C. arenaria*; the former has herbaceous inner faces to the leaf-sheaths. The two species could occur together at the ecotone between slack and dune. Two close relatives occur on the W European dunes: *C. reichenbachii* Bonnet, which has all the spikes ♂ below; and *C. ligerica* Gay, with fewer spikes which are all ♀ at top and ♂ below (or the lowest rarely ♀), and darker, chestnut ♀ glumes. Either may turn up in Britain.

C. arenaria hybridises with *C. remota* but this hybrid has not been recorded for Britain.

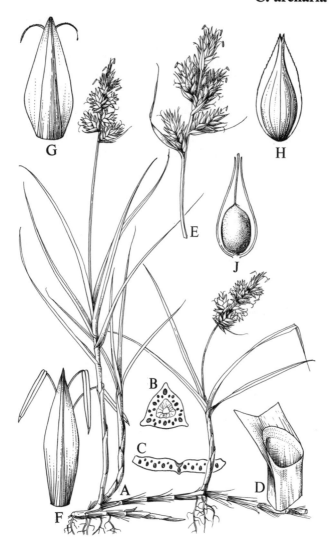

10. Carex disticha Hudson

Brown Sedge Map 13

Rhizomes far-creeping; shoots single or in pairs: roots grey-brown, much branched; scales brown, becoming dark and fibrous. *Stems* 20–100 cm, rough, sharply trigonous. *Lvs* 15–60 cm × 2–4 mm, rough on veins beneath, thick, ± flat but with keeled midrib, gradually tapering to a flat rough tip, mid-green; sheaths forming false stems in sterile shoots, inner face hyaline only around concave apex, otherwise herbaceous, lower sheaths with very short blades, brown, persistent; ligule 3–7 mm, obtuse, tubular. *Infl.* a dense panicle, 2–7 cm, forming a broadly lanceolate to ellipsoid head; bracts glumaceous, brown-hyaline or the lowest lf-like and exceeding infl. *Spikes* numerous, contiguous, sessile, terminal ♀ (sometimes overtopped by ♂ spikelets), intermediate ♂, lower all ♀ or sometimes ♂ at base. ♂ *glumes* 4–5 mm, lanceolate, pale red-brown, with hyaline margins; apex acute. ♀ *glumes* 3.5–4.5 mm, ovate-lanceolate, pale red-brown, with hyaline margins; apex acute. *Utricles* 4–5 mm, ovoid, many-ribbed, red-brown, with very narrow, ± serrate, lateral wings; beak c. 1 mm, rough, split; stigmas 2; nut ovate, biconvex, shortly stalked. Fr. 7–8.

A plant of mixed herb-sedge fens and wet meadows, the single shoots often overlooked if not in flower, preferring areas with a somewhat fluctuating water-table. More frequent in the calcareous fens of the eastern half of England and S Scotland but extending to Caithness; scattered in Ireland but confined to the lowland.

The scattered shoots from a monopodial rhizome distinguish this species from most other *Carex* species. When growing in a dune-slack it may intermingle with *C. arenaria*. The latter species has a ♂ terminal spike and a hyaline inner face to the leaf-sheath.

No hybrids of *C. disticha* are recorded.

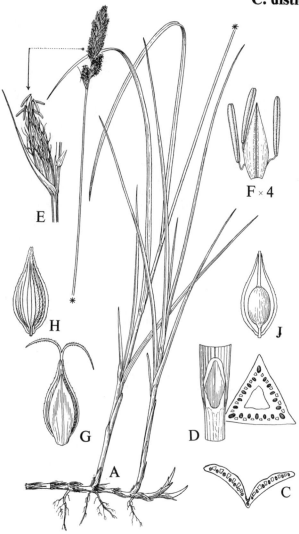

11. Carex chordorrhiza   L.fil.

String Sedge   Map 14

Rhizomes far-creeping, often ascending; shoots solitary, arising from elongated, decumbent base of flowering stem; roots red-brown; scales pale yellow-brown, chaffy, soon decaying. *Stems* 15–40 cm, stout, ± terete, striate, with few short lvs at base. *Lvs* up to 30 cm × 1–2 mm, mid-green, stiff, erect, flat or ± involute, gradually narrowed to a ± trigonous fine point; sheaths forming short false stem, inner face hyaline-brown, apex concave, lower sheaths pale brown-hyaline, often darkening; ligule 1–2 mm, rounded. *Infl.* a compact ± ovoid head, 7–15 mm; bracts glumaceous. *Spikes* 2–4, 4–8 mm, ♂ at top, ♀ at base, lower often entirely ♀. ♂ *glumes* c. 3.5 mm, oblanceolate-elliptic, pale red-brown; apex acute. ♀ *glumes* 3–4 mm, broadly ovate-elliptic, pale red-brown, hyaline towards margin; apex ± acute. *Utricles* 3.5–4.5 mm, ovoid-ellipsoid, ± compressed, yellow- or often dark-brown, shiny, faintly ribbed; beak c. 0.5 mm, bifid; stigmas 2; nut obovoid-oblong, stalked, apex truncate.   Fr. 7–8.

C. chordorrhiza is a plant of very wet base-poor mires with *C. limosa*, *Sphagnum fallax*, *S. papillosum* etc. First found in 1897 by Rev. E. S. Marshall and W. A. Shoolbred at the head of L. Naver, Altnaharra, Sutherland and since found in several places within a three mile radius of there; recently found in similar mires, in quantity, in Easterness, where it was discovered in 1978 by Miss S. E. Page and Dr J. O. Riely. The record for S Uist is in error.

Distinctive in having creeping decumbent shoots which in their second year give rise to similar shoots from the basal four or five, ± widely spaced, nodes. The leaves are often inrolled, the plant having the appearance of marram grass.

A hybrid with *C. curta* (*C.* × *lidii* Flatb.) has been described from Norway.

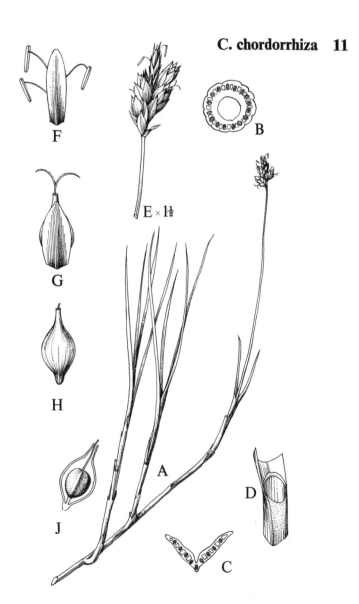

$E \times 1\frac{1}{2}$

12. Carex divisa Hudson

Divided Sedge Map 14

Rhizomes thick, woody, sometimes far-creeping and branched; shoots ± densely clustered, arising from short lateral branches; roots black- or grey-brown; scales pale brown, ribbed, very soon becoming dark and fibrous. *Stems* 15–80 cm, rough at top, wiry, trigonous. *Lvs* 15–60 cm × 1.5–3 mm, stiff, ± flat or often channelled or inrolled, tapering to a slender ± flat tip, mid- to grey-green, overwintering; sheaths of sterile shoots forming a false stem, lower brown, soon decaying, inner face hyaline, apex straight or concave, brown-purple; ligule 2–3 mm, obtuse, ± tubular. *Infl.* 1–3 cm; bracts lf-like or setaceous, lowest usually much exceeding infl. *Spikes* 3–8, contiguous, the lower sometimes remote, 3–13 mm, the upper ♂ at top, ♀ below, lower often all ♀. ♂ *glumes* 3.5–4.5 mm, lanceolate-elliptic, pale red-brown; apex acute. ♀ *glumes* 3.5–5 mm, ovate, purple-brown, with pale midrib and broadly hyaline margins; apex aristate. *Utricles* 3.5–4 mm, ovoid or broadly ellipsoid, plano-convex, faintly nerved, pale brown; beak 0.5–0.75 mm, parallel-sided, bifid; stigmas 2; nut suborbicular, biconvex. Fr. 7–8.

C. divisa is a plant of damp pasture or marshy ground mostly in inorganic soils with *Phragmites, Scirpus maritimus, C. distans, C. disticha, C. nigra* etc; also beside ditches but rarely in the water. It flourishes in brackish conditions and is rarely found far from the coast and may require a high magnesium content of the soil. Mainly in the S and E of England reaching the S Wales coast in Glamorgan and N to the Humber estuary and Holy Island; the two Scottish records are possibly introductions. Not seen recently in Ireland but possibly still in Wexford and elsewhere.

This species has a distinct woody, fibrous rhizome system with the shoots often on short thick side branches. The infl. is distinctly purplish-brown with a characteristic stiff, ± setaceous, lower bract over-topping it.

No hybrids have been confirmed.

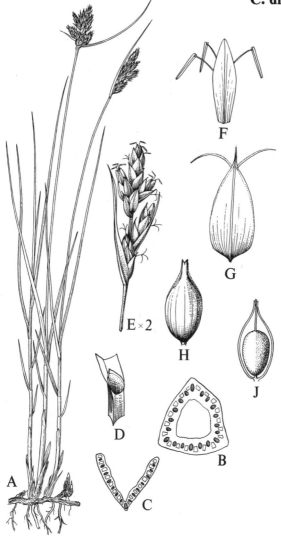

13. **Carex maritima** Gunnerus

Curved Sedge   Map 12

Rhizomes far-creeping, much branched; shoots loosely tufted or solitary and terminal on short branches of rhizome; roots pale red-brown with numerous branched laterals; scales dark brown, soon becoming fibrous. *Stems* 1–18 cm, often curved, terete, striate, solid. *Lvs* 3–15 cm × 0.5–2 mm, thick, stiff, channelled, often inrolled, tapering gradually to a ± trigonous point, mid-green, often overwintering; sheaths hyaline or brown, thin, ribbed, persistent, inner face with thin hyaline membranous strip only (see *Fig.* D), apex straight; ligule 0.5–1 mm, rounded. *Infl.* a compact ovoid or subglobose head, 0.5–1.5 cm; bracts glumaceous or 0. *Spikes* 4–8, clustered, 3–6 mm, few-fld, upper ♂, often hidden, lower ♀ with occasional ♂ at top. ♂ *glumes* c. 4 mm, elliptic, dark orange-brown; apex ± acute. ♀ *glumes* 3–4 mm, broadly ovate, red-brown (often bleached), midrib paler, margins narrowly hyaline; apex ± acute or obtuse and mucronate. *Utricles* 4–4.5 mm, ovoid- subglobose, faintly ribbed, brown, almost black on maturity; beak 0.5–1 mm, bifid; stigmas 2, often persistent; nut orbicular, biconvex.

Fr. 7–8.

C. maritima is a rare plant of sandy coasts, able to withstand salt spray and silt accretion to a limited extent; often found with *C. distans, C. extensa, Plantago maritima, Juncus gerardii,* etc. Recorded for 16 vice-counties around the N and E Scottish coast reaching as far south as Holy Island; also in the outer Hebrides and the northern isles with an isolated station in N Lancs.

This species is perhaps superficially like a *Juncus* sp. when in the vegetative state but the thick leaves are not septate and are inserted in three rows. It is distinguished from all other sedges of its habitat by the narrow hyaline strip on the inner face of the leaf sheath.

Hybrids between *C. maritima* and *C. dioica* have been recorded in N Europe.

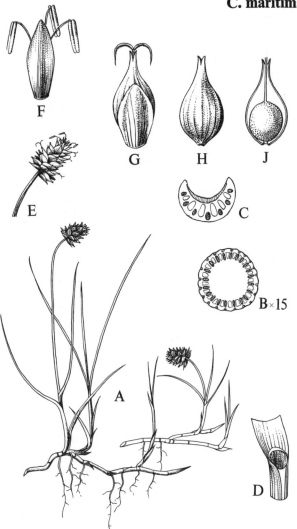

14. Carex remota L.

Remote Sedge   Map 15

Rhizomes short; shoots densely tufted forming stools up to 30 cm high; roots pale purple-brown; scales brown, persistent or more rarely becoming fibrous. *Stems* 30–75 cm, spreading, trigonous or with two serrulate angles towards top. *Lvs* 25–60 cm × 1.5–2 mm, channelled, gradually tapering to a long slender pendulous point, mid-green, overwintering; sheaths pale yellow-brown, persistent, forming false stems in sterile shoots, inner face narrow, hyaline, apex concave; ligule 1–2 mm, rounded or obtuse. *Infl* $\frac{1}{4}$–$\frac{1}{3}$ length of stem; lower bracts lf-like, exceeding infl., upper glumaceous. *Spikes* 4–9, 3–10 mm, sessile, upper ± contiguous, ♀ at top, ♂ at base, lower remote, entirely ♀. ♂ *glumes* 2.5–3 mm, ovate-elliptic, pale brown-hyaline, with green midrib; apex acute. ♀ *glumes* c. 2.5 mm, lanceolate to ovate, pale brown or hyaline, with green midrib; apex acute. *Utricles* 2.5–3.5 mm, ovoid-ellipsoid, plano-convex, green, ± shiny; beak 0.5 mm, broad, split; stigmas 2; nut ovoid, biconvex.   Fr. 7–8.

C. remota is a species of shady situations on peaty or siliceous soils with a high water level for at least part of the year. Commonly found in alder or wet birch carr often with *C. laevigata*, *C. paniculata* etc. Mainly a lowland plant and common throughout England and Wales except in the winter-cold Fenland-Breckland basin. In lowland and W Scotland, absent from extreme N and Outer Hebrides; throughout Ireland, sometimes on calcareous soils, but rarer in the W.

This species is not easily confused with any other; the remote spikes, long bracts and tussocky habit distinguish it.

C. remota hybridises with *C. otrubae* (= *C.* × *pseudaxillaris* K. Richter), *C. paniculata* (= *C.* × *boenninghausiana* Weihe) and *C. divulsa* (= *C.* × *emmae* L. Gross); in all cases the long bract is dominant in the hybrid. Hybrids with *C. arenaria*, *C. curta*, *C. echinata*, *C. elongata* and *C. ovalis* are reported from Europe.

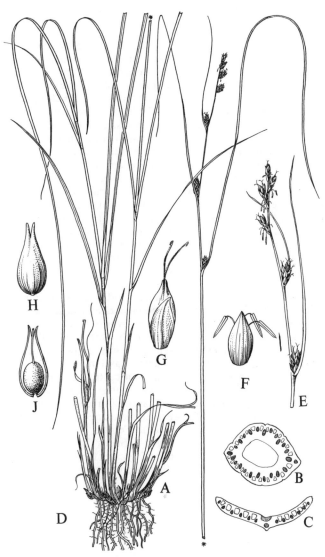

15. Carex ovalis Good.

Oval Sedge Map 16

Rhizomes short; shoots densely tufted, often $\pm$ prostrate; roots pale- or purple-brown; scales dark brown, becoming fibrous. *Stems* 10–90 cm, rough at top, stiff, often curved, trigonous, $\pm$ solid. *Lvs* up to 50 cm × 1–3 mm, thin, $\pm$ soft, margins rough, $\pm$ flat, gradually tapering to a $\pm$ fine trigonous point, mid- to dark-green, often overwintering; sheaths forming a false stem in sterile shoots, becoming pink- or grey-brown, persistent, inner face narrow, hyaline, apex $\pm$ straight; ligule c. 1 mm, obtuse, tubular. *Infl.* a compact ovoid head; lower bracts often setaceous, upper glumaceous. *Spikes* 2–9, contiguous or overlapping, 5–15 mm, sessile, upper ♀ at top, ♂ at base, lower all ♀. ♂ *glumes* 4–5 mm, broadly lanceolate, pale orange-brown with keeled midrib and broadly hyaline margin; apex acute. ♀ *glumes* 3–4.5 mm, lanceolate-elliptic, dark or red-brown, with green or paler midrib and narrowly hyaline margin; apex acute. *Utricles* 4–5 mm, ellipsoid-ovoid, plano-convex, narrowly winged at top, distinctly nerved, light brown; beak c. 1 mm, winged, rough, bifid; stigmas 2; nut obovoid or oblong-ellipsoid, biconvex, shortly stalked. Fr. 7–8.

C. ovalis is a species of moderately acid soils with impeded drainage. Found in wet meadows, woodland rides, rough heathland where water accumulates and in *Festuca-Agrostis-Nardus* grasslands of upland Britain. Common throughout the British Isles although local in the E Anglian fens and less frequent in S central Ireland.

The prostrate shoots with elongated basal nodes and coarse leaves, which often form tufts with open centres, are distinctive. When in fruit, the $\pm$ ovoid heads with rather divaricate utricles, giving a 'spiky' appearance to the infl., are easily recognisable.

C. ovalis hybridises with *C. remota* in Germany and Switzerland. A N American taxon, *C. crawfordii* Fernald, very similar in appearance but differing in having a narrowly lanceolate utricle longer than the acute and lanceolate ♀ glumes (Figs G_2, H_2), has in the past become established with introduced crop seed.

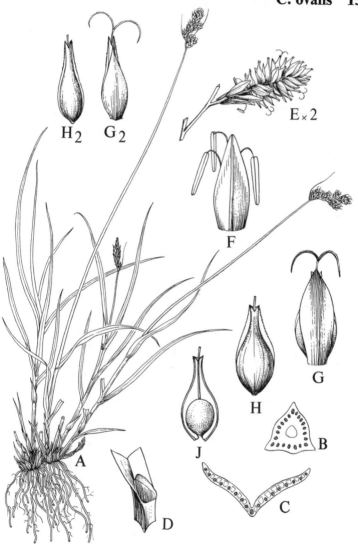

16. Carex echinata Murray

Star Sedge Map 17

Rhizomes very short; shoots densely tufted; roots whitish; scales pale brown, persistent. *Stems* 10–40 cm, slender, trigonous and ± striate to subterete. *Lvs* up to 30 cm × 1–2.5 mm, thick, keeled or becoming flat, gradually tapered to a trigonous rough tip, shiny, mid- to yellow-green; sheaths often white with green veins, becoming pale brown, soon decaying, inner face hyaline-green, apex ± straight; ligule c. 1 mm, rounded, tubular. *Infl.* 1–3 mm; bracts glumaceous or rarely setaceous and equalling the infl. *Spikes* 2–5, ± distant, 3–6 mm, sessile, terminal spike ♀ at top, ♂ below, lower spikes all ♀. ♂ *glumes* 2.5–3 mm, broadly lanceolate, pale brown with broad hyaline margin; apex obtuse. ♀ *glumes* 2–2.5 mm, broadly ovate, embracing lower part of utricle, pale red-brown, with green midrib and broad, hyaline margin; apex acute. *Utricles* 3–4 mm, ovoid, plano-convex, faintly ribbed, green becoming yellow-brown, divaricate at maturity; beak c. 1 mm, broad, serrate, bifid; stigmas 2; nut obovoid or orbicular-compressed.   Fr. 6–8.

C. echinata is a species of mesotrophic soils which are seasonally or permanently waterlogged, e.g. heath reverting to bog with *Erica tetralix, Scirpus cespitosus, Eriophorum vaginatum*, etc. In oligotrophic mires usually within the pH range of 4.5–5.7, e.g. *Carex-Molinia* or alpine *Carex-Sphagnum* flushes from 90–750 m (300–2500 ft). Also in more eutrophic mires with silty, not peaty, soils with a high base status, such as those over limestone. Abundant in the N and W of the British Isles and on the wet, sandy heaths of the S, SE and E; rarer in the Midlands due to drainage.

Easily distinguished by the squarrose mature utricles which form a characteristic star-shaped spike; in the young stages the few-fld spikes are the best distinction. The glossy, thick, dark green leaves with whitish sheaths with green veins, help to separate it from other Carices of its habitat.

A hybrid with *C. curta* (*C.* × *biharica* Simonkai) is recorded from Scotland and that with *C. dioica* (= *C.* × *gaudiniana* Guthn.) from Ireland; hybrids with *C. paniculata* and *C. spicata* are said to occur but have not been reported from Britain.

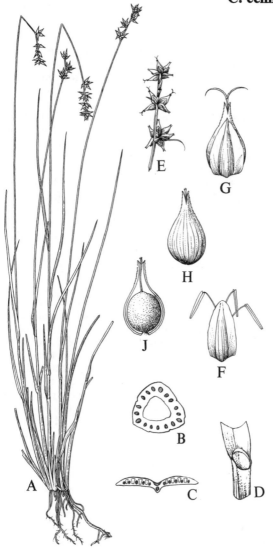

17. Carex dioica   L.

Dioecious Sedge   Map 18

Plants dioecious. *Rhizomes* shortly creeping; shoots loosely tufted, often decumbent; roots often as thick as rhizome, pale purple-brown; scales pale purple- or orange-brown, soon decaying or becoming fibrous. *Stems* 5–30 cm, erect, terete, striate. *Lvs* 5–20 cm × 0.3–1 mm, with 3 veins, rigid, ± erect, channelled, and incurved; apex rounded, dark green, dying to a dull red-brown; sheaths becoming pale orange-brown, persistent, inner face hyaline, apex straight but soon split; ligule 0.5 mm, rounded, tubular. *Infl.* a single terminal, usually unisexual, spike; bracts glumaceous or 0. ♂ *spike* 8–20 mm; ♂ *glumes* 3–4 mm, ovate-oblong, red-brown with hyaline margin; apex obtuse or acute. ♀ *spike* 5–20 mm, subglobose to cylindric; ♀ *glumes* 2.5–3.5 mm, red- to purple-brown, with pale midrib, and dark nerve and ± hyaline margin, persistent; apex obtuse or acute; that of the lowest floret often enlarged, with an acuminate apex. *Utricles* 2.5–3.5 mm, patent or rarely deflexed when ripe, broadly ovoid, compressed, pale red- to purple-brown, with darker ribs; beak 0.5–0.75 mm, serrulate, notched; stigmas 2; nut subglobose, compressed.   Fr. 7–8.

  C. dioica is a species of eutrophic mires in wet, silty muds, rarely in pure peat, in pH range 5.5–6.5. Found in calcareous flushes and springs at various altitudes, with *C. demissa*, *flacca*, *nigra*, *panicea* and *pulicaris*, *Campylium stellatum* etc. Common on the mica schists and limestones of N and W Scotland; likewise in N Wales, N Pennines and Lake District. Scattered in lowland Britain and Ireland, but rapidly diminishing due to drainage.
  Variable in size of ♀ spike, utricle and whole plant. The narrow leaf with three veins is a useful vegetative character. Occasionally found with a few ♀ fls at base of the ♂ spike and then may be mistaken for *C. pulicaris*. Hybrids with *C. echinata* (= *C.* × *gaudiniana* Guthn.) have been reported from Ireland; also with *C. curta*, *C. lachenalii* and *C. maritima*, from Europe.
  C. parallela (Laest.) Sommerf., a N Scandinavian plant with narrower utricles (Fig. H²) with a smooth beak, has been confused with forms of *C. dioica*; it has not been confirmed for Britain but could possibly occur. Another dioecious species, the Central European *C. davalliana* Sm., with long-beaked, oblong-lanceolate utricles (Fig. H³) was once found in Somerset (see p. 250); other British records for this species have proved to be forms of *C. dioica*.

100

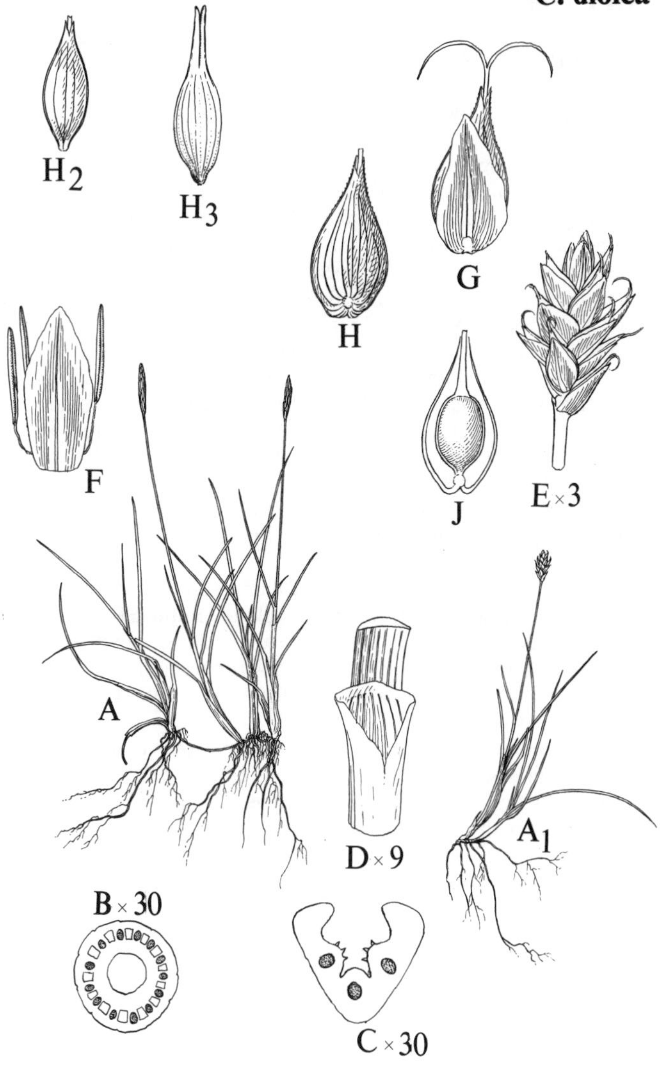
H2
H3
G
H
F
E×3
J
A
D×9
A1
B×30
C×30

18. Carex elongata L.

Elongated Sedge   Map 19

Rhizomes short; shoots densely tufted; roots pale brown; scales grey-brown, usually persistent. *Stems* 30–80 cm, rough with upwards pointing teeth, trigonous. *Lvs* 25–90 cm × 2–5 mm, rough beneath, thin, ± flat or slightly keeled, tapering gradually to a very fine, flat tip, mid-green, often red-brown and persistent on dying; sheaths forming short false stems, lower pale or pink-brown, shiny, persistent, inner face hyaline, apex concave; ligule 4–8 mm, acute, with little or no free margin. *Infl.* 3–7 cm, lax; lower bracts setaceous, upper glumaceous. *Spikes* 5–18, ± contiguous, 5–15 mm, ± erect, becoming divaricate on ripening, sessile, upper ♀ at top, ♂ at base, lower entirely ♀. ♂ *glumes* 2.5–3 mm, ovate-oblong, pale red-brown, with green midrib and broadly hyaline margin; apex rounded or obtuse, ± mucronate, ♀ *glumes* c. 2 mm, ovate-elliptic, red-brown, midrib green; apex acute or obtuse. *Utricles* 3–4 mm, lanceolate-ellipsoid, plano-convex, often curved, distinctly ribbed, green becoming dark brown; beak 0.5–0.75 mm, often minutely serrulate, truncate; stigmas 2; nut compressed-cylindric, stalked.   Fr. 6–7.

C. elongata is a very local plant of damp soil in water meadows, beside ditches and often in boggy woodland; in the SE as far W as the Hants–Dorset border, and in Warwickshire; scattered from Denbigh and Cheshire to Yorks and Cumberland and north to Loch Lomond. In Ireland around Lough Neagh, and in Fermanagh, Cavan and Roscommon.

This is a distinct species characterised by the long, divaricate, brownish spikes and the narrow, ellipsoid, ribbed utricle; vegetatively the long ligule and pink-brown persistent non-fibrous sheaths separate it from the *C. muricata* group.

Hybrids with *C. appropinquata*, *C. paniculata* and *C. remota* are recorded from Europe but not from the British Isles.

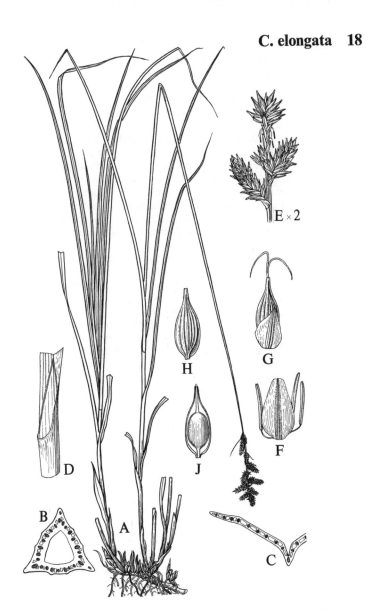

E × 2

19. Carex lachenalii Schkuhr

Hare's-foot Sedge   Map 19

Rhizomes shortly creeping; shoots closely tufted; roots brown;
scales brown, soon fibrous. *Stems* up to 20 (–30) cm, bluntly
trigonous, striate. *Lvs* shorter than stems, 1–2 mm, broad, flat,
dark green; ligule c. 1 mm, rounded. *Infl.* 2–4 mm, red-brown.
Spikes 2–5, closely contiguous, the upper at the apex of a triangle
completed by the next two below, 5–8 mm, subclavate, ♀ at top, ♂ at
base. ♂ *glumes* 2–3 mm, elliptic, apex obtuse, red brown. ♀ *glumes*
c. 2.5 mm, ovate, red-brown, with broad hyaline margin, apex obtuse.
Utricles 3–4 mm, ovoid, markedly narrowed at base, usually green
below and brown above; beak c. 0.5 mm, smooth, split at the back,
the halves overlapping; stigmas 2; nut obovoid, biconvex. Fr. 7–9.

A very local but sometimes abundant plant of wet slopes and rock
ledges at 750–1140 m (2500–3800 ft) on acid mountains in the
Cairngorms. Also known in one station in the Ben Nevis range and
one in Glencoe. A record from Harris is almost certainly erroneous,
for this is a plant of high snow-patches.

Closest in morphology to *C. ovalis*, but in its British habitats
cannot be confused with any other sedge except its hybrid with *C.
curta* (*C.* × *helvola*), which grows with the parents in some abundance
on Lochnagar and on Cairntoul and has been reported from the
Clova mountains. The hybrid has the larger, more contiguous
spikes of *C. lachenalii* with the pale colouring of *C. curta*. It is
completely sterile.

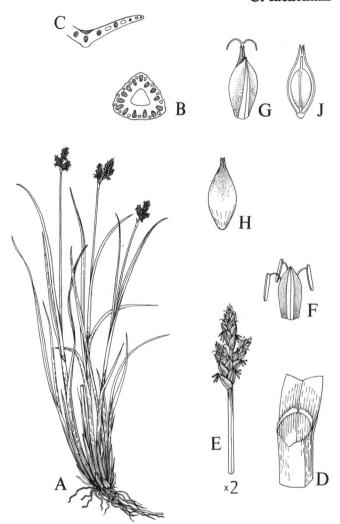

20. Carex curta Good.

White Sedge Map 20

Rhizomes very shortly creeping; shoots loosely tufted; roots pale brown or whitish; scales pale or pink-brown, usually persistent. *Stems* 10–50 cm, rough above, slender, sharply trigonous. *Lvs* 15–55 cm × 2–3 mm, soft, thin, flat or ± keeled, tapering ± gradually to a fine, rough, ± flat tip, pale green; sheaths thin, forming short false stems only, lower pink-brown, soon decaying, inner face hyaline, apex ± straight; ligule 2–3 mm, acute. *Infl.* 3–5 cm; bracts glumaceous, lower sometimes setaceous. *Spikes* 4–8, contiguous or ± distant, 5–8 mm, ♀ at top, ♂ at base. *♂ glumes* 2–2.5 mm, ovate or broadly elliptic, hyaline, with green midrib; apex obtuse. *♀ glumes* c. 2 mm, ovate-oblong, hyaline, with green midrib; apex acute or apiculate. *Utricles* 2–3 mm, ovoid-ellipsoid, plano-convex, pale or blue-green to yellow, with yellowish ribs; beak 0.5–0.75 mm, minutely rough, notched; stigmas 2; nut obovoid or ellipsoid, biconvex. Fr. 6–8.

C. curta is a component of the oligotrophic alpine *Carex-Sphagnum* mires with species such as *C. aquatilis*, *C. rariflora* and *C. nigra* around 720 m (2400 ft). Found also in lowland Britain in more mesotrophic mires and wet places, usually associated with *C. rostrata*, *Eriophorum angustifolium* etc. On wet sandy heaths in SE England, becoming frequent in NW Wales, N England and Scotland; in Ireland rare in the S and W.

The soft pale green leaves and the acutely trigonous stems with pale fruiting heads distinguish this from any other sedge.

Hybrids with *C. echinata* (= *C.* × *biharica* Simonkai) and *C. lachenalii* (= *C.* × *helvola* Blytt ex Fries) are found in Scotland. Hybrids with *C. appropinquata*, *C. curta*, *C. dioica*, *C. elongata*, *C. paniculata* and *C. remota* are reported from elsewhere.

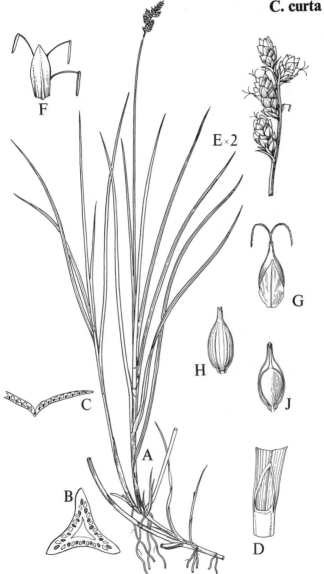

21. Carex hirta L.

Hairy Sedge   **Map 21**

Rhizomes often far-creeping; shoots tufted; roots pale brown, often much branched; scales brown or red-brown, eventually becoming fibrous. *Stems* 15–70 cm, trigonous with rounded faces. *Lvs* 10–50 cm × 2–5 mm, ± hairy on both surfaces, flat or ± keeled, gradually tapering to a fine point, mid-green, often overwintering; sheaths hairy, those of sterile shoots forming a rigid false stem, inner face hyaline, often densely hairy, apex ± straight; ligule 1–2 mm, obtuse, free margin fringed with hairs. *Infl.* up to $\frac{3}{4}$ length of stem; lower bracts lf-like, longer than spike but not exceeding infl., upper setaceous. ♂ *spikes* 2–3, 10–25 mm; ♂ *glumes* 4–5 mm, obovate-oblanceolate, ± hairy, red-brown with pale midrib or often ± hyaline throughout; apex mucronate. ♀ *spikes* 2–3, contiguous or ± distant, 10–45 mm, cylindric, erect; peduncles smooth, up to twice as long as spike, half-ensheathed; ♀ *glumes* 6–8 mm, ovate-oblong, green-hyaline, midrib excurrent as a green ciliate awn. *Utricles* 5–7 mm, ovoid, many ribbed, green, hairy; beak 2 mm, rough, hairy, deeply bifid; stigmas 3; nut obovoid, trigonous, stalked.   Fr. 6–9.

C. hirta is a plant of various soils, usually in grassy associations such as hedge-banks, meadows, consolidated sand dunes etc, usually in hollows or channels where soil moisture accumulates; occasionally in damp woods. Throughout Britian, becoming less common in Scotland and very rare north of the Great Glen; markedly absent from areas with a rainfall over 1000 mm (40 ins) per year. Scattered throughout Ireland except in blanket bog areas.

This species can be confused with little else and is distinctive with its hairy utricles, ♂ glumes and lvs. In specimens growing in damp shady situations the pubescence is lost from the leaves and sheaths *except* near the apex of the hyaline inner face; the utricles and glumes are always hairy. This form has been called *C. hirta* var. *sublaevis* Hornem. *C. hirta* hybridises with *C. vesicaria* (= *C.* × *grossii* Fiek) and with *C. rostrata* but the latter hybrid has not been recorded for the British Isles.

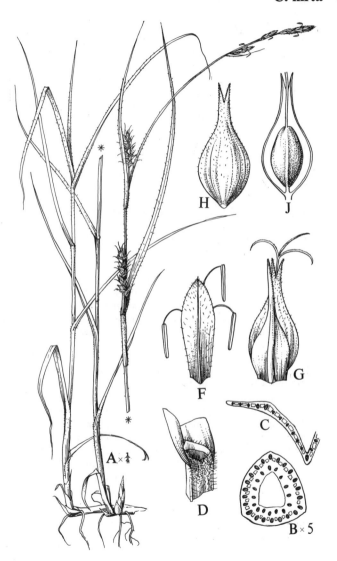

A × ¼

H

J

F

G

D

C

B × 5

22. Carex lasiocarpa Ehrh.

Slender Sedge Map 22

Rhizomes far creeping; shoots loosely tufted, slender; roots pale yellow-brown; scales pale or yellow-brown, occasionally wine-red. *Stems* 45–120 cm, slender, stiff, smooth or slightly rough above, trigonous, striate. *Lvs* 30–100 cm × 1–2 mm, grey-green, stiff, flat but usually inrolled, long-attenuate into a fine acicular, whip-like point; sheaths dark purple-brown to pale red-brown, soon decaying, inner face membranous, purple- or often pale brown, persistent or often fibrillose, apex straight, dark purplish; ligule 2–3 mm, obtuse. *Infl.* $\frac{1}{8}$–$\frac{1}{6}$ length of stem; bracts lf-like, very slender, lower often exceeding infl. ♂ *spikes* 1–3, 20–70 mm; ♂ *glumes* 4–6 mm, lanceolate, purple-brown, with green or pale midrib; apex acute. ♀ *spikes* 1–3, contiguous or ± distant, 10–30 mm, cylindric-oblong, erect, ± sessile; ♀ *glumes* 3.5–5 mm, lanceolate, chestnut-brown, with pale midrib; apex acute or acuminate. *Utricles* 3.5–4.5 mm, ovoid, tomentose, grey-green; beak 0.5–1 mm, ± deeply bifid; stigmas 3; nut ovoid, trigonous. Fr. 7–9.

C. lasiocarpa is a species of mesotrophic to eutrophic mires and reed swamp, on substrata ranging from sedge peats to sandy raw soils of exposed lake shores. It sometimes forms pure stands in submerged situations but is more often associated with *Myrica gale*, *C. rostrata* and *Sphagnum subsecundum*; this association develops into a *C. lasiocarpa-Myrica* scrub vegetation with much *Sphagnum fallax*. In Ireland, it is associated with *Phragmites, C. rostrata, C. limosa, Eriocaulon* etc; often associated with *Cladium* and hypnoid mosses in lowland Britain. Scattered throughout the British Isles.

The narrow grey-green lvs with red- or purple-brown lower sheaths distinguish this species from any other in these habitats; in fruit it is easily recognisable by its tomentose utricles. For vegetative differences between *C. buxbaumii* and *C. lasiocarpa* see p. 192.

In Britain *C. lasiocarpa* forms a hybrid with *C. riparia* (= *C.* × *evoluta* Hartman); hybrids with *C. rostrata* and *C. vesicaria* are reported from Europe but not from the British Isles.

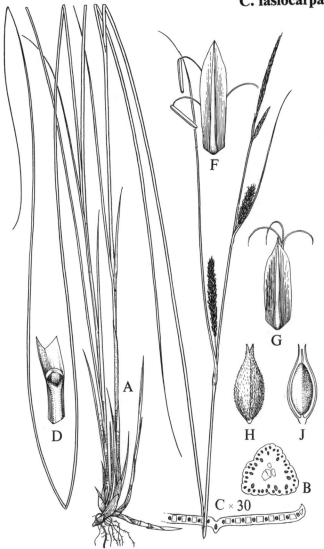

23. Carex acutiformis Ehrh.

Lesser Pond Sedge

Map 23

Rhizomes wide creeping; shoots tufted; roots brown, often dark; scales grey-brown, soon becoming fibrous. *Stems* 60–150 cm, rough, often smooth below, sharply trigonous, solid. *Lvs* up to 160 cm × 7–10 mm, thin, keeled or plicate, arcuate, gradually tapering to apex, glaucous at first, becoming dull, often red-green at apex, red-brown on dying; sheaths brown, usually red-streaked, persistent inner sheaths not translucent, inner face hyaline-brown, persistent, usually fibrillose on splitting, apex concave; ligule 5–15 mm, acute. *Infl.* c. ⅓ length of stem; bracts lf-like, exceeding infl. ♂ *spikes* 2–3, clustered, 1–4 cm, lower with setaceous bracts; ♂ *glumes* 5–6 mm, oblong to oblanceolate, purple-brown with a pale midrib; apex obtuse or subacute. ♀ *spikes* 3–4, ± contiguous, 2–5 cm, cylindric, erect, upper sessile often ♂ at top, lowest only shortly pedunculate; ♀ *glumes* 4–5 mm, oblong-lanceolate, red- or purple-brown with paler midrib; apex acute or acuminate. *Utricles* 3.5–5 mm, ellipsoid-ovoid, ribbed, greyish-green; beak c. 0.5 mm, notched; stigmas 3, rarely 2; nut obovoid, trigonous, apex flat.   Fr. 7–9.

Similar to *C. riparia* in its ecological preferences and often growing with that species, *C. acuta* and *C. elata* in swamps, semi-swamp-carr and wet meadows in lowland Britain and Ireland. Similar in distribution to *C. riparia*, but more frequent than that species in W and N England; rare in Scotland with its most northerly locality in Banffshire.

This species, if robust, may be confused with depauperate *C. riparia*. The obtuse or subacute male glumes which are never mucronate are usually sufficient to distinguish it, if an inflorescence is available. Vegetatively the inner brownish opaque sheaths seldom show the transverse septa and veins as clearly as *C. riparia*. Both *C. acuta* and *C. elata* have a narrower, more glaucous, leaf; *C. elata* has buff-coloured basal scales and no long rhizomes; *C. acuta* is less distinct but has a rounder stem base and brown, persistent, inner faces to the sheaths which lack any fibrillae. Both the latter species have two stigmas.

C. acutiformis hybridises in Britain with *C. riparia*, and *C. acuta* (= *C.* × *subgracilis* Druce). The first two need confirmation. Hybrids with *C. elata*, *C. flacca*, *C. lasiocarpa* and *C. vesicaria* are recorded for Europe.

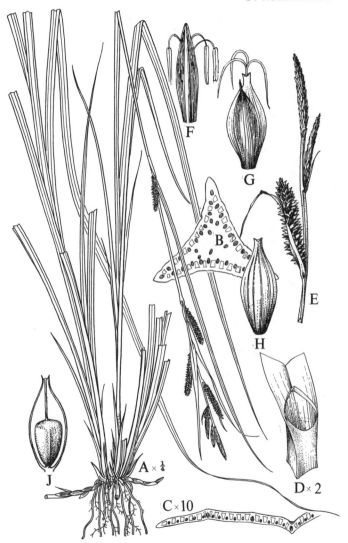

24. Carex riparia   Curtis

Greater Pond Sedge   Map 24

Rhizomes far-creeping; shoots tufted; roots thick, pale brown; scales grey-brown, soon becoming fibrous. *Stems* 60–130 cm, rough, sharply trigonous, solid. *Lvs* up to 160 cm × 6–15 mm, rigid, erect, thin, sharply keeled or plicate, abruptly attenuate to short trigonous apex, glaucous, persisting as pale brown litter; sheaths persistent, grey-brown or often red-tinged, white-hyaline at base with distinct transverse septa; inner face hyaline, becoming brown and persisting, rarely forming fibrillae on splitting, apex concave; ligule 5–10 mm, obtuse or rounded. *Infl.* c. $\frac{1}{3}$ length of stem; lower bracts lf-like, exceeding infl., upper setaceous. ♂ *spikes* 3–6, contiguous, 2–6 cm; ♂ *glumes* 7–9 mm, oblong-lanceolate, dark brown with midrib and margins paler; apex acuminate. ♀ *spikes* 1–5, ± contiguous, 3–10 mm, cylindric-fusiform, upper erect, ± sessile, often ♂ at top, lower pedunculate; peduncles rough, often as long as spike, shortly ensheathed; ♀ *glumes* 7–10 mm, oblong-lanceolate or narrowly ovate, dark, often purple-brown, midrib paler or green; apex acuminate. *Utricle* 5–8 mm, ovoid, inflated, green or brown, apex tapered; beak 1.5 mm, bifid; stigmas 3; nut oblong-ovoid, trigonous, stalked.   Fr. 6–9.

A plant often forming large stands by slow flowing rivers, in ditches and around ponds, sometimes with *C. acutiformis*, *C. acuta*, *C. pseudocyperus* etc. In some situations it may replace *C. acutiformis*; in others it may be entirely absent. The ecological preferences of the two are not clear-cut. *C. riparia* forms a dominant (or co-dominant with *C. acutiformis*) herb layer in wet fen woods (carr) of E Anglia; it often forms large colonies where water stands for long periods in Spring. Mainly a lowland plant and therefore more frequent in S and E England; in Scotland not found N of Argyll and Aberdeenshire and in Ireland almost confined to S and E.

The colour and erect habit of the lvs is sufficient to distinguish this species from *C. acuta* or *C. acutiformis* when *en masse*; when mixed with the latter species in a fen the more translucent inner lf-sheaths and usual lack of fibrillae and abruptly tapered lf-tip may help to separate *C. riparia*.

In Britain *C. riparia* hybridises with *lasiocarpa* (= *C.* × *evoluta* Hartman), *rostrata* (= *C.* × *beckmanniana* Figert) and *vesicaria* (= *C.* × *csomadensis* Simonkai); the second needs confirmation. Hybrids with *C. elata* and *C. flacca* are recorded for Europe.

114

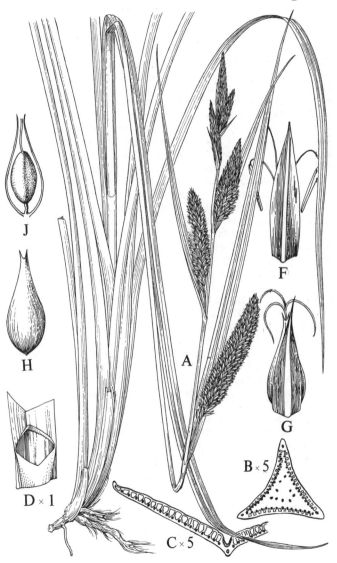

J

H

D × 1

A

F

G

B × 5

C × 5

25. Carex pseudocyperus L.

Cyperus Sedge   Map 25

Rhizomes short; shoots loosely tufted; roots thick (up to 1.5 mm), orange-brown, felted; scales dark- or grey-brown, persistent. *Stems* 40–90 cm, solid, sharply trigonous, angles rough. *Lvs* up to 120 cm × 5–12 mm, rigid, erect, thin, plicate, very rough on keel and margins, tapered gradually to a fine point, bright yellow-green, passing through yellow to grey-brown on dying, persistent; sheaths becoming pink- or grey-brown, persistent, inner face hyaline, forming fibrillae on splitting, apex straight or concave; ligule 10–15 mm, obtuse. *Infl.* $\frac{1}{6}$–$\frac{1}{4}$ length of stem; bracts lf-like, lowest 3–4 times longer than infl. ♂ *spikes* 1, 20–60mm; ♂ *glumes* 5–7 mm, elliptic-lanceolate, brown with a green or paler midrib; apex long-acuminate, ciliate. ♀ *spikes* 3–5, clustered at top and often exceeding the ♂ spike, 20–100 mm, cylindric, pendulous; peduncles slender, rough, lowest only shortly ensheathed; ♀ *glumes* 5–10 mm, ovate, brownish hyaline, with green midrib, drawn out into a long, fine, ciliate arista. *Utricles* 4–5 mm, broader than glumes, ovoid-ellipsoid, ribbed, green, patent, soon falling when ripe; beak c. 2 mm, smooth, deeply bifid; stigmas 3; nut obovoid, trigonous.   Fr. 7–8.

A plant of eutrophic or mesotrophic open water swamps; often along sides of slow-flowing dykes and in ponds, oxbows and derelict canals. It can tolerate some shade and is occasionally found in pools in woods. Predominantly a lowland plant. Mainly in the SE and Midlands extending as far north as N Lancs; rare in Wales and one isolated locality in NE Scotland (Moray); scattered but less frequent than formerly in Ireland.

This species is unmistakable and stands out amongst other riparian sedges by its yellow-green colour and pendulous, bristly ♀ spikes which soon lose their utricles at maturity.

C. pseudocyperus forms sterile hybrids with *C. rostrata* (*C.* × *justi-schmidtii* Junge) and *C. vesicaria*; only the former is recorded for the British Isles (Norfolk).

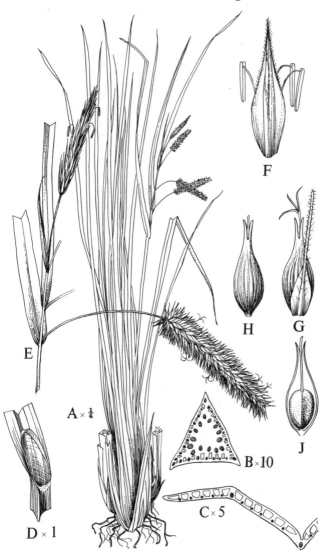

A × ¼

B × 10

C × 5

D × 1

E

F

G

H

J

26. **Carex rostrata** Stokes

Bottle Sedge Map 27

Rhizomes far-creeping; shoots few in each tuft; roots thick, purple-
or orange-brown; scales pale grey-brown, rarely red-tinged, soon
decaying. *Stems* 20–100 cm, subterete and smooth below, trigonous
and rough above. *Lvs* 30–120 cm × 2–7 mm, rough, rigid, keeled or
plicate, or in some habitats inrolled, tapering to a long (2–6 cm)
acicular point, glaucous on upper surface, dark green and shiny
beneath, overwintering; sheaths herbaceous, thick, ± spongy, dark
brown, often streaked with red, inner sheaths pink, inner face
hyaline, becoming brown and often fibrillose on splitting, apex
straight; ligule 2–3 mm, rounded. *Infl.* up to ½ length of stem;
bracts usually lf-like, equalling or exceeding infl. ♂ *spikes* 2–4,
20–70 mm, lower ones with setaceous bracts; ♂ *glumes* 5–6 mm,
elliptic-oblanceolate, brown with a paler midrib; apex acute or
obtuse. ♀ *spikes* 2–5, contiguous or lowermost distant, 30–80 mm,
cylindric, suberect, subsessile or lowest shortly stalked; ♀ *glumes*
3–5.5 mm, narrower than utricle, oblong-lanceolate, purplish-brown
with pale midrib; apex acute. *Utricle* 3.5–6.5 mm, ovoid, inflated,
faintly ribbed, yellow-green, patent; beak 1–1.5 mm, smooth, bifid;
stigmas 3; nut subglobose-trigonous. Fr. 7–9.

A plant of swamps, lake-margins and peaty areas with high water
level where the pH is between 4.5 and 6.5; also occurs as a flush
plant where base status is not too high. A common and important
component of upland mesotrophic wetlands throughout Scotland,
Ireland and the N of England; in the S it is scattered in the more
acid fens.

C. rostrata is confused with *C. vesicaria* which can be distinguished
by the longer, acute ligule, the lack of spongy lf-sheaths and the
regular fibrillae, and the narrower, longer, more tapered utricle.

There is considerable minor variation in fruit and leaf characters
which, in relation to habitat, would repay further study. *C. rhyncho-
physa* Fischer, a closely allied European species, has been recorded in
error in Ireland.

C. rostrata forms hybrids with *C pseudocyperus* (= *C.* × *justi-
schmidtii* Junge), *C. saxatilis* (? *C.* × *ewingii* E. S. Marshall) and
C. vesicaria (*C.* × *involuta* (Bab.) Syme), the latter often forming
large stands; those with *C. saxatilis* need further investigation.
Further hybrids with *C. acutiformis*, *C. hirta*, *C. lasiocarpa* and
C. riparia are reported from Europe.

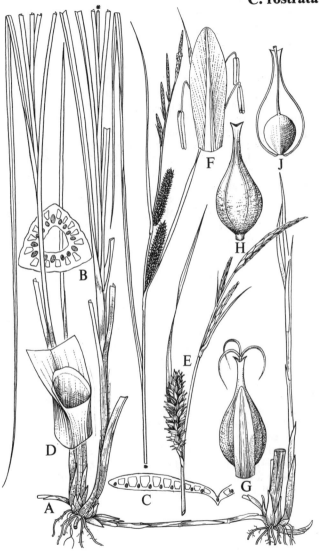

27. Carex vesicaria L.

Bladder Sedge   Map 28

Rhizomes shortly creeping; shoots slender, markedly trigonous, 2–3 in each tuft; roots thick, pale yellow-brown; scales brown- or purple-red, usually persistent. *Stems* 30–120 cm, trigonous, rough on the angles above, smooth below. *Lvs* up to 150 cm × 4–8 mm, serrulate for entire length, thin, rigid, plicate, gradually tapered to a fine point, mid- or yellow-green, quickly decaying; sheaths becoming purple-red, persistent, inner face hyaline, fibrillose on splitting, apex straight or concave; ligule 5–8 mm, acute. *Infl.* $\frac{1}{4}$–$\frac{1}{3}$ length of stem; bracts lf-like, lower exceeding infl. ♂ *spikes* 2–4, 10–40 mm, the lower often with setaceous bracts; ♂ *glumes* 4–6 mm, elliptic or oblanceolate, purplish-brown with green or paler midrib and hyaline margins; apex ± acute. ♀ *spikes* 2–3, ± contiguous but distant from the ♂ spikes, 2–4 cm, oblong-cylindric, erect, subsessile or lowest with peduncle as long as spike; ♀ *glumes* 4–6 mm, narrowly lanceolate, purplish-brown with paler or green midrib; apex acute or acuminate, hyaline. *Utricles* 6–8 mm, ovoid-ellipsoid, inflated, ribbed, ± shiny, olive green, ascending; beak c. 2 mm, smooth, bifid; stigmas 3; nut obovoid, trigonous.   *Fr.* 7–8.

A plant of wet peat-lands but more mesotrophic than *C. rostrata* forming a characteristic community around Scottish lochs with *C. nigra*, *C. disticha*, *Juncus acutiflorus*, *J. effusus* etc. Also in more open situations and on inorganic soils, at edges of streams, dykes and canals (often persisting when colonisation to scrub takes place). Scattered throughout Britain and Ireland to Shetland, although uncommon N of the Great Glen.

Variable in size of utricle and ♀ glume and in leaf structure. Differences from *C. rostrata* are given on p. 118; when sterile the red fibrillose lf-sheaths of the angled, slender shoots of *C. vesicaria* are sufficient to distinguish it.

It hybridises with *C. hirta* (= *C.* × *grossii* Fiek), *C. riparia* (= *C.* × *csomadensis* Simonkai) and *C. rostrata* (= *C.* × *involuta* (Bab.) Syme). For a discussion on hybrids with *C. saxatilis* see p. 122. A hybrid with *C. pseudocyperus* is recorded for Europe.

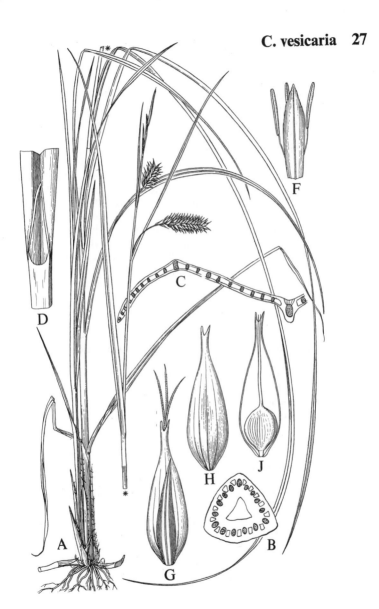

28. Carex saxatilis L.

Russet Sedge   Map 29

Rhizomes far-creeping, producing tufts of 1–3 shoots at frequent, ± regular intervals; roots pale yellow-brown; scales grey-brown often tinged wine-red, soon becoming fibrous. *Stems* 15–40 cm, ± rough above, trigonous, often curved. *Lvs* 12–40 cm × 2–4 mm, often curved, thick, bluntly keeled or channelled, gradually tapering to a trigonous point up to 5 cm long, mid-green, ± shiny, becoming straw-coloured; sheaths thick, white, tinged with wine-red, persistent, inner face hyaline, apex straight; ligule 2–4 mm, rounded. *Infl.* $\frac{1}{6}-\frac{1}{4}$ length of stem; lower bracts lf-like, shorter than or equalling the infl., upper usually glume-like. ♂ *spike* 1, rarely 2, 10–15 mm; ♂ *glumes* 3–4 mm, oblanceolate, purple-black with hyaline margin; apex ± acute. ♀ *spikes* 1–3, ± contiguous, 5–20 mm, ovoid or subglobose, erect, only the lowest rarely shortly pedunculate; ♀ *glumes* 2–3 mm, ovate, purple- to dark red-brown with a paler midrib and hyaline margin; apex ± acute. *Utricles* 3–3.5 mm, ovoid, ± inflated, often dark purple-green on exposed faces, shiny; beak 0.5 mm, ± notched; stigmas 2; nut subglobose. Fr. 8–9.

A plant of higher mountains usually between 750–960 m (2500–3200 ft) altitude. Found in mires where the water movement is not great; tolerant of wide range of pH and Ca^{++} content, expecially frequent on saddles or slightly sloping hillsides with *C. curta, C. nigra, C. echinata, Eriophorum* etc. Confined to the higher Scottish mountains, mainly in the high rainfall areas of the west and usually in places where snow lies late; very local in the Cairngorms and Clova mountains.

A variable plant especially in size, colour and texture of utricle. Reported as having two or three stigmas on same spike; all such plants should be investigated for (partially fertile) hybridity with either *C. rostrata* (= ? *C.* × *ewingii* E. S. Marshall) or *C. vesicaria* (see p. 124). *C. saxatilis* var. *alpigena* Fries is such a form with narrower female glumes and utricles similar to *C. rostrata* and possibly a hybrid with that species. One form, of scattered occurrence, with more olive-green utricles, was described erroneously as a hybrid with *C. flava* (*C. demissa* or *C. lepidocarpa* was intended) by Arthur Bennett and called *C.* × *marshallii* Ar. Benn. (cf. discussion on *C.* × *grahamii* p. 124).

122

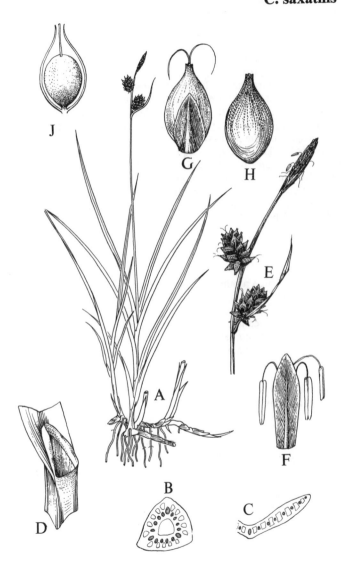

29. Carex × grahamii Boott

Map 30

Rhizomes far-creeping, producing tufts of 2–3 shoots at frequent, ±
regular intervals; roots pale yellow-brown often darkening on
drying; scales purplish or wine-red, soon decaying. *Stems* 25–50 cm,
± rough above, trigonous, often curved. *Lvs* 20–40 cm × 2–4 mm,
thinner and more sharply keeled than in *C. saxatilis*, tapering to a
trigonous point, mid-green, dull, becoming straw-coloured or
grey-brown; sheaths often dark purplish or wine red, persistent,
inner face ± hyaline, apex straight; ligule 3–6 mm, acute. *Infl.*
$\frac{1}{4}-\frac{1}{5}$ length of stem, lowermost bract leaf-like, often with its
subtended spike abortive, ± equalling the infl., upper setaceous or
glume-like. ♂ *spikes* 1–2, rarely 3, 10–25 mm; ♂ *glumes* 4–5 mm,
oblanceolate, purple- or orange-brown; apex subacute. ♀ *spikes* 1–3,
± contiguous, 10–30 mm, cylindrical or ovoid, erect, the lower
pedunculate; ♀ *glumes* 2–3 mm, ovate, red-brown with a paler
midrib and hyaline margin; apex subacute and broadly hyaline.
Utricles 4–5 mm, ovoid, ± tapered to a short (0.5 mm) beak,
translucent, greenish-brown, distinctly ribbed; stigmas 3; nut not
forming.   Fr. 7–8.

A plant of high mountain flushes between 750–900 m (2500–3000
ft) found in inorganic mires on steep slopes or broad ledges,
mainly on the schistose soils of the Breadalbanes with an outlier in
Clova. In the Breadalbanes often associated with, or in the vicinity
of, *C. saxatilis*. Colonies are compact, spreading vegetatively, widely
separated one from another, and often show striking morphological
differences, the Clova plants being much more robust than the
western ones.

In all the material seen the utricles are empty suggesting the
plants are of hybrid origin. Jermy and Tutin (1968) give the putative
parents on morphology alone as *C. saxatilis* and *C. vesicaria*, but
Wallace (1975) points out that the latter has not been found near
C. × grahamii, whereas *C. rostrata*, another putative parent, has.
SEM studies of leaf epidermis indicate that *C. × grahamii* is closer
to *C. vesicaria* than *C. rostrata*, and very close to the Scandinavian
C. stenolepis Less., a fertile species possibly of similar hybrid origin.

C. × grahamii is most easily confused with forms of *C. saxatilis*
which differs in having ovate ♀ glumes 2–3 mm long and more ovoid
ribless utricles consistently bearing 2 stigmas; the leaves of *C.
saxatilis* are more fleshy (i.e. thicker).

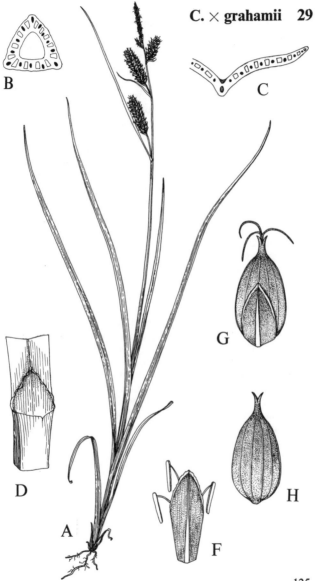

30. **Carex pendula** Hudson

Pendulous Sedge   Map 26

Rhizomes short; shoots in tufts often up to 70 cm across; roots up to 3 mm thick, red-brown; scales red-brown, persistent or becoming fibrous. *Stems* 60–180 cm, trigonous. *Lvs* 20–100 cm × 15–20 mm, rigid, thin, keeled, ± flat, abruptly tapering to a blunt point, margins rough, yellow-green above, ± glaucous beneath, red-brown on dying, often overwintering; sheaths red-brown, persistent, inner face hyaline only at concave apex; ligule 30–60 mm, acute. *Infl.* c. $\frac{1}{3}$ length of stem; bracts lf-like, ± equalling or slightly shorter than infl. ♂ *spikes* 1–2, 6–10 cm; ♂ *glumes* 6–8 mm, lanceolate, brownish hyaline; apex acuminate. ♀ *spikes* 4–5, ± contiguous, lowest often distant, 7–16 cm, cylindric, erect at first becoming pendulous; peduncles rough, ensheathed, lowest half as long as spike; ♀ *glumes* 2–2.5 mm, ovate, red-brown with pale midrib; apex acute or acuminate. *Utricles* 3–3.5 mm, broadly ellipsoid or ovoid, trigonous, ± glaucous-green becoming brown; beak c. 0.3 mm, truncate; stigmas 3; nut obovoid, trigonous.   Fr. 6–7.

C. pendula is a species preferring acid, but base-rich, heavy soils and is frequent in hazel-ash-oak woods and scrubland on Wealden and Oxford clays. Found also on less clayey soils where there is a constant water supply, e.g., in runnels and beside wet ditches; often with *C. remota* and *C. strigosa*. Predominantly a lowland species in S and E England, scattered in Wales and Ireland and scarce and mainly coastal in Scotland, not recorded north of the Great Glen. Frequently an introduction.

A species easily recognisable by its large size. Young plants may be confused with *Scirpus sylvaticus* or with *C. strigosa*, but the red-brown sheaths and long ligule are sufficient to separate *C. pendula*.

No hybrids are known.

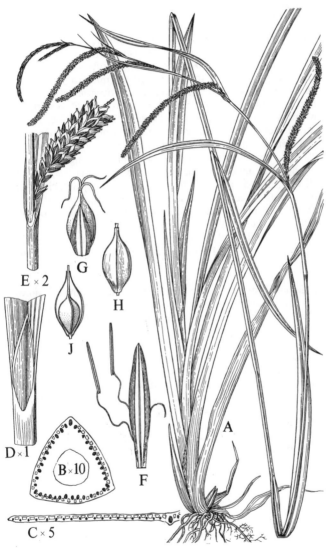

127

31. Carex sylvatica Hudson

Wood Sedge

Map 31

Rhizomes very short; shoots often densely tufted; roots grey-brown; scales brown, soon fibrous. *Stems* 15–60 cm, spreading or nodding, slender, trigonous. *Lvs* 5–60 cm × 3–6 mm, soft, slightly keeled or plicate, abruptly tapered to a sharp point, mid- to yellow-green, becoming brown then bleached on dying, overwintering; sheaths hyaline becoming brown, inner face splitting and persisting as brown membrane, apex concave; ligule c. 2 mm, obtuse. *Infl.* $\frac{1}{3}$–$\frac{1}{2}$ length of stem; lowermost bracts lf-like, sometimes longer than infl., upper setaceous and shorter than spike. ♂ *spike* usually 1, 10–40 mm, very slender; ♂ *glumes* 4–5 mm, oblong-oblanceolate, brown-hyaline; apex acute, or obtuse and mucronate. ♀ *spikes* 3–5, ± distant, 20–65 mm, lax-flowered, pendent; peduncles rough, filiform, up to three times the length of the spike, base ensheathed; ♀ *glumes* 3–5 mm, ovate-lanceolate, hyaline, straw-coloured or brown, with green midrib; apex acute or acuminate. *Utricles* 3–5 mm, ellipsoid- or obovoid-trigonous, green, with two prominent lateral nerves; beak 1–1.5 mm, bifid; sitgmas 3; nut ellipsoid, ± trigonous.   Fr. 4–8.

A plant of heavy often wet soils in woods although sometimes on chalky soils with little clay; occasionally in open scrub and grassland but then more likely a relic of woodland and doubtfully established in the open. Frequent throughout Britain and Ireland but uncommon in the main Grampian massif and rare in N Scotland.

This species can be confused with little else in this habitat except *C. strigosa*, which has utricles with a very short beak (c. 0.2 mm). Non-fruiting plants can be distinguished by the ligule, which in *C. sylvatica* is c. 2 mm, obtuse, and with a very short (0.5 mm) free portion; in *C. strigosa* the ligule is c. 5 mm, acute and with at least 1 mm of free tissue.

A hybrid with *C. strigosa* has been recorded in France and with *C. hirta* in Austria.

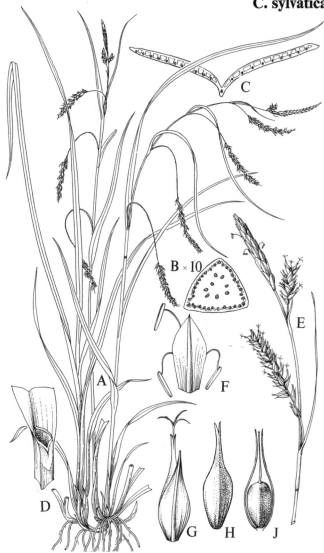

32. Carex capillaris L.

Hair Sedge   Map 32

Rhizomes short; shoots in ± open tufts; roots slender, purple-brown; scales dark- or red-brown, soon becoming fibrous. *Stems* 10–40 cm, bluntly trigonous to subterete. *Lvs* 5–10 cm × 0.5–2.5 mm, stiff, sometimes arcuate, flat or ± channelled, gradually tapered to a short subulate tip (*Fig.* K), grey-green, changing to red- then grey-brown on dying, usually overwintering; sheaths red-brown, soon becoming fibrous, inner face hyaline, apex straight; ligule c. 1 mm, rounded. *Infl.* up to $\frac{1}{2}$ length of stem; bracts lf-like, usually a single one subtending a cluster of spikes, ± exceeding infl., that of distant spike shorter. ♂ *spike* 1, 5–10 mm, few-flowered, overtopped by ♀ spikes; ♂ *glumes* 2–3 mm, oblong-obovate, hyaline with a brown midrib; apex obtuse to rounded. ♀ *spikes* 2–4, 5–25 mm, rarely more than 10-fld, clustered and appearing to arise from a single node, lowest sometimes distant; peduncles up to 4 cm, hair-like, half ensheathed; ♀ *glumes* 2–3 mm, caducous, broadly ovate, hyaline or straw-coloured; apex acute or mucronate. *Utricles* c. 3 mm, narrowly ovoid-ellipsoid, olive to dark brown, smooth, shiny; beak 0.5 mm, truncate; stigmas 3; nut ellipsoid, trigonous.
Fr. 7–8.

A plant of wet hillsides, often with *Festuca ovina*, *Agrostis* spp., *C. flacca* etc, mostly on base-rich soils or areas flushed by base-rich water; it can tolerate some shade. Found also in mineral-rich bogs, where specimens may become luxuriant. Rare in Snowdonia, local in N Pennines and S Uplands, and frequent in the Grampians and in limestone areas of N Scotland where it descends to sea level.

A species not likely to be confused with any other when flowering. In the vegetative state it can be difficult to separate from *C. pilulifera* which may grow in adjacent grassland not flushed with base-rich water. The latter tends to form larger tufts with wider lvs which have a longer acicular point and are often appressed to the ground.

No hybrids are known.

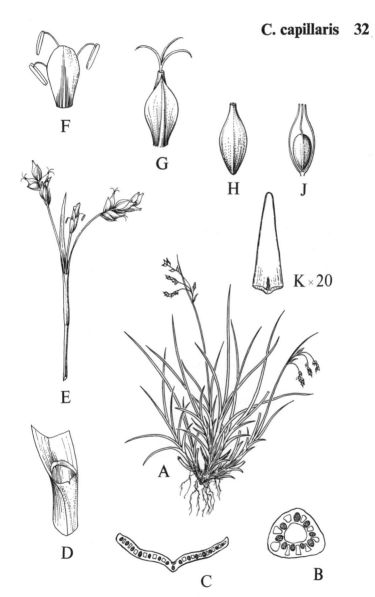

K × 20

33. Carex strigosa Hudson

Thin-spiked Wood Sedge Map 29

Rhizomes short; shoots tufted; roots pale brown; scales orange- or red-brown, becoming fibrous. *Stems* 35–70 cm, trigonous-subterete, often spreading. *Lvs* 15–40 cm × 6–10 mm, thin, ± arcuate, plicate, with two lateral veins prominent on upper surface, abruptly tapered to sharp point, mid-green, pink-brown on dying; sheaths thin, brown, occasionally red-tinged, persistent, inner face hyaline, apex concave; ligule 5–8 mm, acute. *Infl.* $\frac{1}{2}$–$\frac{3}{4}$ length of stem; bracts lf-like, longer than spikes, not exceeding infl. ♂ *spike* 1, 30–40 mm; ♂ *glumes* 4.5–5.5 mm, narrowly obovate, brown, with green midrib; apex acuminate. ♀ *spikes* 3–6, distant, lowest often remote, 25–80 mm, lax-flowered, ± erect; peduncles smooth, half ensheathed; ♀ *glumes* c. 2.5 mm, ovate or ovate-lanceolate, green becoming brown; apex acute. *Utricles* 3–4 mm, oblong- or narrowly ellipsoid, often curved, green; beak c. 0.3 mm, truncate; stigmas 3; nut subglobose, trigonous, shortly stalked. Fr. 6–9.

C. strigosa is a plant of base-rich loamy or, less frequently, heavy soils especially near streams or in damp hollows; usually in open glades in hazel-oak-alder woodland. Most frequent in the lowland wooded areas of the Weald, Somerset and Severn valley and scattered throughout England, becoming rare in the E and N, as far as E Yorks; in scattered localities in N and E Ireland.

It grows in similar situations to *C. sylvatica* and is most often confused with that when sterile; when in fruit the rounded utricle with a very short beak easily distinguishes it. The lvs of *C. strigosa* are usually wider and not so soft as *C. sylvatica* and the ligule in the latter is shorter (2–3 mm), obtuse and with a very narrow (c. 0.5 mm) free portion. Small *C. pendula* has red-brown scales and lf-sheaths tinged with wine-red, and a longer ligule than *C. strigosa*.

A hybrid with *C. sylvatica* has been recorded in France.

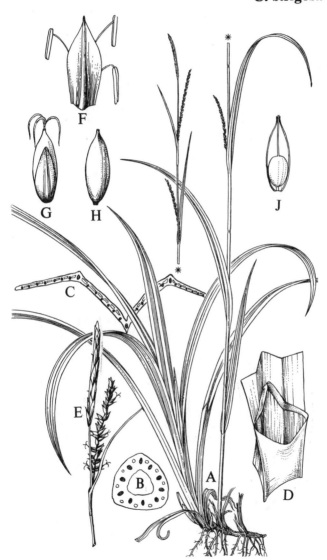

34. Carex flacca Schreber

Glaucous Sedge Map 35

Rhizomes often far-creeping; shoots loosely tufted; roots pale grey-brown; scales dark-brown or red-brown, persistent. *Stems* 10–60 cm, rigid, trigonous-subterete. *Lvs* up to 50 cm × 1.5–4 mm, rigid, often arcuate, flat, gradually tapering to a fine point, glaucous beneath (best seen when lf folded in shoot) dark, dull green above, persisting as dark- or rich-brown litter; sheaths becoming dark-brown, often wine-red, persistent, entire, inner face hyaline-brown, apex straight or concave; ligule 2–3 mm, rounded, shortly tubular. *Infl.* $\frac{1}{5}$–$\frac{1}{3}$ length of stem; bracts lf-like, lowest slightly exceeding infl. ♂ *spikes* (1–) 2–3, 10–35 mm; ♂ *glumes* 3–4 mm, oblanceolate, purple-brown, with pale midrib and hyaline margin; apex rounded or subacute. ♀ *spikes* 1–5, 15–55 mm, contiguous, cylindric, upper erect, subsessile, often ♂ at top, lower ± nodding; peduncles rough, as long as spike, half ensheathed; ♀ *glumes* 2–3 mm, oblong-ovate, purple-black, pruinose, with a wide paler midrib and hyaline margin; apex obtuse, mucronate. *Utricle* 2–3 mm, broadly ellipsoid to obovoid, often inflated on adaxial side, minutely papillose, yellow-green, often turning deep purple-black; beak 0.2 mm, truncate; stigmas 3; nut ellipsoid, trigonous. Fr. 6–9.

C. flacca is the commonest sedge of calcareous grassland; found also on sand-dunes with *Festuca rubra* etc, in calcicolous dwarf shrub, e.g. *Dryas*, communities, and on boulder clay. It can withstand some salinity and is often found in the estuarine marshes. Also a component of eutrophic flushes with *C. demissa*, *C. nigra*, *C. panicea* etc. Possibly the most ubiquitous sedge in the British flora.

This is a variable species in stature, length of spikes and coloration of utricles and glumes. Forms with dark glumes are mistaken for *C. nigra*, but that species has only two stigmas and a flattened, biconvex utricle. Further the lvs of *C. flacca* are green above and glaucous only beneath. *C. nigra* has lvs which are equally glaucous on both sides; the lvs of *C. panicea* which are even more glaucous on both sides have a distinctive subulate tip. Plants which have been recorded as hybrids of *C. flacca* and *C. nigra* (= *C.* × *winkelmannii* Ascherson & Graebner) have usually been found to be forms of the latter. Hybrids with *C. tomentosa*, *C. montana*, *C. nigra* and *C. panicea* have been recorded for Europe.

134

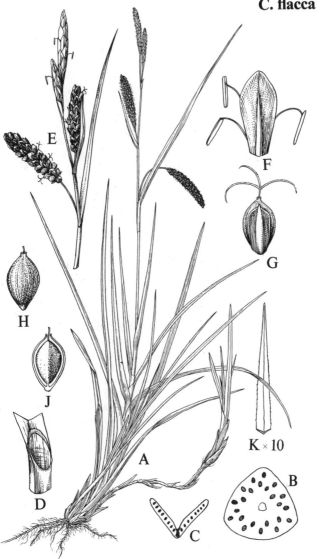

35. Carex panicea L.

Carnation Sedge Map 33

Rhizomes shortly creeping; shoots tufted; roots pale grey- or purple-brown; scales grey-brown, soon decaying. *Stems* 10–60 cm, often curved above, trigonous-subterete, striate, with basal lvs as long as stem. *Lvs* up to 60 cm × 1.5–5 mm, rough at top, ± flat, tapering to a trigonous point (*Fig.* K), glaucous, pale straw on dying; sheaths not loose, white or pale pink-brown, ribbed, persistent but fibrous on decay, inner face hyaline, soon decaying, apex straight; ligule 1.5–2 mm, obtuse. *Infl.* $\frac{1}{6}$–$\frac{1}{4}$ length of stem; bracts lf-like, not loose-sheathed, 1–2 times length of spike. ♂ *spike* 1, 10–20 mm; ♂ *glumes* 3–4.5 mm, ovate-oblong to elliptic, purple-brown, midrib pale, margin hyaline; apex ± obtuse. ♀ *spikes* 1–3, 10–15 mm, ± distant, few and lax-fld; ♀ *glumes* 3–4 mm, broadly ovate, purple or red-brown, mibrib pale, margin hyaline; apex acute. *Utricles* 3–4 mm, broadly obovoid, inflated on adaxial side so apex points outwards (see H[2]), olive-green or ± purple-tinged; beak less than 0.5 mm, truncate; stigmas 3; nut oblong-obovoid, trigonous. Fr. 6–9.

    C. panicea is found on soils of pH 4–7 and of varying base content, being less frequent on those not receiving ± continuous irrigation. It occurs in mountain grassland; dwarf-shrub heaths on deep peat; oligotrophic mires with *Molinia–C. echinata*; meso- and eutrophic mires with other sedges and brown mosses; and in alpine flushes with *C. nigra*, *C. pulicaris* etc. Throughout British Isles. A species on which morphological and ecological studies are needed.

    C. panicea, when sterile, can be confused with *C. flacca* and *C. nigra* but neither of these two species has the trigonous lf-tip. Hybrids with *C. magellanica* and *C. vaginata* are recorded in Scandinavia but not from Britain.

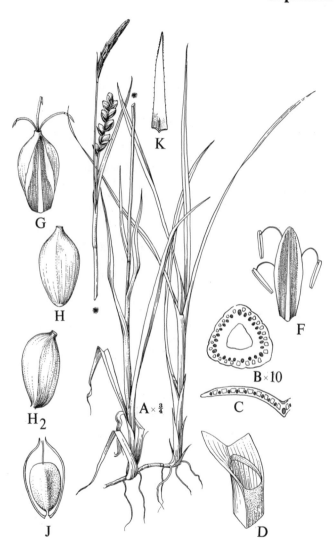

36. **Carex vaginata** Tausch

Sheathed Sedge Map 34

Rhizomes far creeping; shoots tufted; roots brown; scales grey-brown, fibrous. *Stems* 15–40 cm, trigonous-subterete, striate. *Lvs* up to 60 cm × 3–6 mm, keeled, parallel-sided for much of their length and rather abruptly tapered to a flat point, yellow- or bronzy-green, basal decumbent, those on the stem with loose sheaths and subulate blades 1–3 cm long; ligule very short, truncate. *Infl.* ¼ length of stem; lower bracts loosely sheathing (funnel-shaped) and shorter than spike. ♂ *spike* 1, 10–15 mm, ± clavate; ♂ *glumes* more than 4.5 mm, orange-brown, with a paler midrib. ♀ *spikes* 1–2 (–3), 10–20 mm, ± distant, erect, few- and lax-flowered; ♀ *glumes* 3 mm, broadly ovate, red-brown with paler midrib, apex acute or mucronate. *Utricles* c. 4 mm, broadly obovoid, hardly inflated, ribbed, with an obliquely truncate beak c. 1 mm long; stigmas 3; nut ovoid-trigonous. Fr. 7–9.

An alpine plant occurring above 600 m (2000 ft) on wet rocky sills and a characteristic constituent of flushed grassland in Breadalbane and the Cairngorms. Often abundant but usually shy-flowering and therefore liable to be overlooked. Very local in Dumfriesshire and Roxburghshire, one station in W Ross and one in W Sutherland.

Can only be confused with *C. panicea*, from which the yellow (not glaucous) colouring, the flat leaf-tip, and the markedly loose sheaths of the stem-leaves and bracts at once distinguish it.

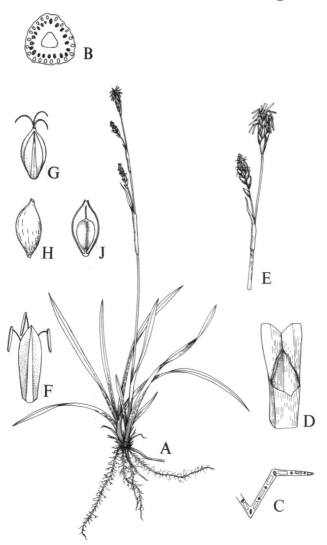

37. Carex depauperata   Curtis ex With.

Starved Wood Sedge   Map 30

Rhizomes shortly creeping; shoots few, loosely tufted; roots pale purple-brown; scales purple- to red-brown, shiny, persistent. *Stems* 30–100 cm, stout, trigonous-subterete. *Lvs* 20–60 cm × 2–4 mm, thin, flat, tapering from a broad base to a fine point, mid- to yellow-green, dying to a pale straw; sheaths purplish- or red-brown, shiny, persistent, inner face hyaline, flecked with red, becoming brown, apex straight or concave; ligule 2–3 mm, obtuse. *Infl.* $\frac{1}{4}$ length of stem; bracts lf-like, longer than spikes, upper sometimes exceeding infl. ♂ *spike* 1, 18–30 mm; ♂ *glumes* 5–6.5 mm, elliptic, (lowest on spike up to 10 mm, bract-like), pale or red-brown with a paler or green midrib and hyaline margins; apex ± acute. ♀ *spikes* 2–4, distant, 2–6-fld, 10–20 mm, erect; peduncles rough, half-ensheathed; ♀ *glumes* 4.5–6 mm, broadly lanceolate to obovate, brown with a green midrib and broad hyaline margins; apex acute or mucronate. *Utricle* 7–9 mm, rhomboid-obovoid, narrowed into a solid base, distinctly ribbed, brownish-green, shiny; beak 3 mm, smooth or scabrid, obliquely truncate, split in front; stigmas 3; nut obovoid, trigonous.   Fr. 6–7.

A very rare plant of dry woods and hedgebanks on chalky or limestone soils. In Surrey, N Somerset, Anglesey and Mid Cork but recently seen only in the second and fourth of these. Formerly in Kent (now part of Greater London); one unconfirmed record from near Wimborne, Dorset.

Easily recognisable by its large utricles and few-flowered spikes. Vegetatively similar to *C. tomentosa* which, although of restricted range, could occur in similar habitats. The latter has fibrillose lf-sheaths and ± glaucous leaves.

No hybrids are known.

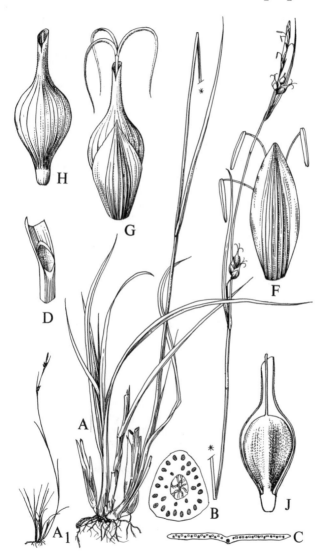

38. Carex laevigata   Sm.

Smooth-stalked Sedge   Map 36

Rhizomes short; shoots forming dense tufts up to 30 cm across; roots yellow-brown, felty; scales brown or reddish, rarely becoming fibrous. *Stems* 30–120 cm, stout, trigonous with slightly rounded faces. *Lvs* 15–60 cm × 5–10 mm, smooth, thin, shallowly keeled or plicate, abruptly tapered, bright green, brown and persisting when dead; sheaths persistent, brown, inner face hyaline, apex convex or lingulate; ligule 7–15 mm, obtuse. *Infl.* $\frac{1}{4}-\frac{2}{3}$ length of stem; bracts lf-like, longer than spike but not exceeding infl. ♂ *spikes* 1 or 2, 20–60 mm; ♂ *glumes* 5–6 mm, oblong-oblanceolate, pale orange-brown, margins and base hyaline; apex obtuse and often mucronate, less often long-acute. ♀ *spikes* 2–4, distant, 20–50 mm, ovoid-cylindric, ± erect but lowest usually pendent; peduncles up to 80 mm, ensheathed; ♀ *glumes* 3–5 mm, ovate or ovate-lanceolate, brown with green midrib which is often rough at tip; apex acuminate. *Utricles* 4–6 mm, ovoid or subglobose, partially inflated, ± strongly ribbed, green with fine reddish dots; beak 1.5 mm, ± scabrid, deeply bifid; stigmas 3; nut trigonous-globose, shortly stalked.   Fr. 7–8.

C. laevigata is a plant of shady moist situations, more rarely in open habitats; most common in woodlands on acid, but base-rich, clay soils. Frequent in Britain and Ireland where annual rainfall is over 750 mm (30 ins) i.e. the Weald, SW and W Britain and SE and W Ireland.

This species could be confused initially and in marginal habitats with either *C. distans* or *C. binervis* but the fine red dots on the utricles and the acuminate ♀ glumes which lack a hyaline edge in *C. laevigata* are distinctive. Further the ♂ glumes of *C. binervis* and *C. distans* are purplish, not gingery-brown. Vegetatively the differences are not so obvious; the dark green lvs of *C. binervis* have splotches of wine-red on dying, which are never seen on *C. laevigata*, and the much shorter ligule and usually narrower leaf of *C. distans* are sufficient to distinguish that species.

Hybrids with *C. binervis* (= *C.* × *deserta* Merino), *C. demissa* and *C. pallescens* have been found in Britain.

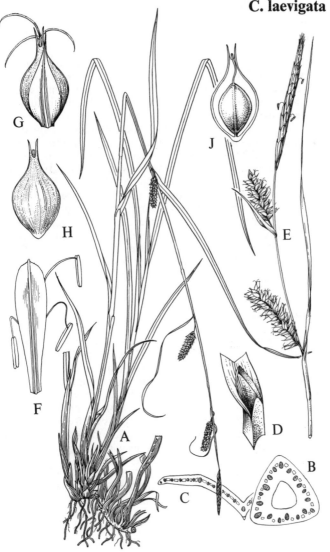

39. Carex binervis Sm.

Green-ribbed Sedge   Map 38

Rhizomes shortly creeping; shoots forming dense clumps in wet habitats, less tufted in closed grasslands; roots thick, grey-brown; scales orange-brown, persistent, *Stems* 15–150 cm, trigonous-terete, often with a single furrow. *Lvs* 7–30 cm × 2–6 mm, rigid, often arcuate in dwarf plants, keeled or ± flat, ± abruptly tapering to a fine point, dark green, dull, with wine-red splotches on dying, persisting as pink- or orange-brown litter, frequently overwintering; sheaths dull, red-brown, persistent, inner face hyaline, apex lingulate at least on stem-lvs (*Fig.* D^1), convex or straight (*Fig.* D) on lower lvs; ligule 1–2 mm, rounded. *Infl.* up to ½ length of stem; lower bracts lf-like, 2–4 times as long as the spike, upper glume-like. ♂ *spike* 1, 20–45 mm; ♂ *glumes* 4–4.5 mm, oblong-obovate, purplish with paler midrib; apex obtuse or rounded and mucronate, erose, scarious. ♀ *spikes* 2–4, distant, 15–45 mm, cylindric, erect, lowermost usually nodding; peduncles half-ensheathed, lowest up to 10 cm; ♀ *glumes* 3–4 mm, ovate, dark purple-brown with green or pale brown midrib; apex obtuse, mucronate. *Utricle* 3.5–4.5 mm, purple-brown or rarely green, broadly elliptic, with two prominent green lateral ribs; beak 1–1.5 mm, rough, bifid; stigmas 3; nut obovoid, trigonous, olive-brown.   Fr. 6–8.

Predominantly a plant of acid, siliceous soils most common in mountain grasslands of *Agrostis* spp.–*Molinia* and in *C. bigelowii*–*Diphasiastrum alpinum* turf; less frequently on peaty soils but seen also on wet alpine cliff ledges. In lowland situations found on sandy heaths, rough pastures and rocky cliffs. Frequent throughout Britain and Ireland but less so in the drier S Midlands and E Anglia.

It is difficult to distinguish in the vegetative state from *C. bigelowii* but lacks the purplish-brown scales and the glaucous lvs of that species; further *C. bigelowii* never has the wine-red splotches on the dying lvs. In maritime situations *C. binervis* may be confused with *C. distans* (see p. 146); *C. sadleri* E. F. Linton differs only in having narrower utricles with a longer beak and less distinct lateral ribs.

C. binervis hybridises with *C. demissa* (= *C.* × *corstophinei* Druce), *C. laevigata* (= *C.* × *deserta* Merino) and *C. punctata* in Britain. Hybrids with *C. flava* and *C. lepidocarpa* are recorded from Europe.

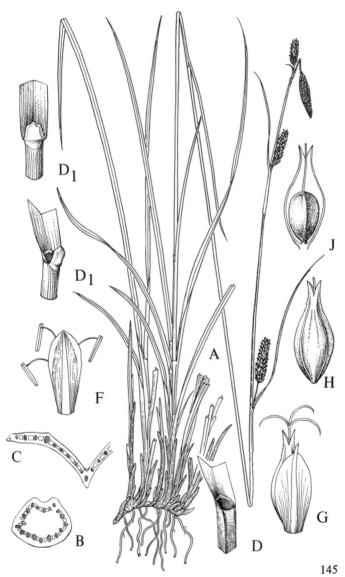

40. Carex distans L.

Distant Sedge Map 39

Rhizomes short; shoots $\pm$ densely tufted; roots red-brown; scales dark brown or black, rarely wine-red. *Stems* 15–100 cm, smooth, trigonous-terete. *Lvs* 10–15 cm × 2–6 mm, rigid, $\pm$ erect, flat, tapered to a fine point, grey-green rapidly becoming brown, then ash-grey and persisting on dying, overwintering; sheaths persisting, eventually becoming fibrous, dark to mid-brown, the younger ones orange-brown, inner face herbaceous, hyaline towards top with a brown margin, apex convex or straight (*Fig.* D) or in upper leaves, protruding (*Fig.* D¹); ligule 2–3 mm, obtuse. *Infl.* compact at flowering, elongating on fruiting to at least $\frac{2}{3}$ length of stem; bracts lf-like, the lower shorter than the adjacent internode, the upper longer but not exceeding stem. ♂ *spike* usually 1, 15–30 mm; ♂ *glumes* 3–4 mm, obovate, pale to purplish brown; apex subacute to obtuse-mucronate. ♀ *spikes* 2–4, 10–20 mm, oblong-cylindric, erect; peduncles up to 40 mm, ensheathed; ♀ *glumes* 2.5–3.5 mm, ovate-oblong, brown rarely chestnut, with greenish midrib and hyaline margins; apex acute to obtuse, mucronate. *Utricle* 3.5–4.5 mm, trigonous-ellipsoid, rounded at base, tapered at apex, green rarely dark brown, distinctly nerved, inserted at an angle of 45–60° to the stem-axis; beak 0.75 mm, $\pm$ rough, bifid; stigmas 3; nut trigonous-ellipsoid, yellow-brown. Fr. 6–7.

Throughout the British Isles but not recorded for Shetland. North of the Severn–Wash line and in Ireland a coastal plant of rocky or sandy places within the spray zone or on beaches in reach of the higher spring tides; also in brackish marshes, when it is less tufted. In the S and E it is found in inland mineral-rich marshes where perhaps magnesium is a necessary requirement.

Resembles *C. punctata* (see p. 148), *C. hostiana* and *C. binervis*. *C. hostiana* is easily identified by the trigonous blunt tip to the leaf. In *C. binervis* the utricles are darker, often mottled purplish and the nut is olive-brown, flattened at the top; ♀ glumes are usually a purple-brown and the spikes pendulous; the rhizome scales are uniformly red-brown or wine-red and the decaying leaves blotched with wine-red remain a pink-brown and do not grey as in *C. distans*. *C. distans* hybridises with *C. extensa* (= *C.* × *tornabenii* Chiov.), *C. hostiana* (= *C.* × *muellerana* F. W. Schultz) and *C. lepidocarpa*. Hybrids with *C. flava* and *C. serotina* have been recorded in Europe.

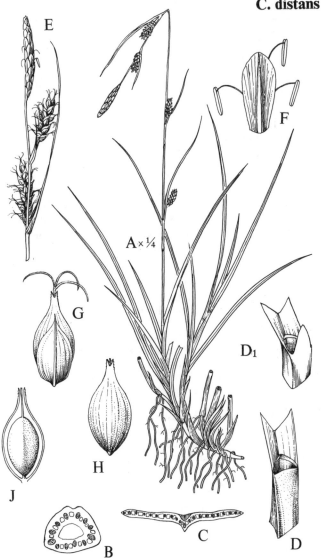

41. Carex punctata Gaudin

Dotted Sedge   Map 40

Rhizomes shortly creeping; shoots tufted; roots orange-brown to black, not felty; scales brown, rarely red, becoming fibrous. *Stems* 15–100 cm, trigonous. *Lvs* 10–50 cm × 2–5 mm, usually as long as the stem, but variable, flat or shallowly keeled, abruptly tapered to a fine tip, pale or yellow-green, persisting on dying as a grey-brown litter, doubtfully overwintering; sheaths persistent, orange- or pink-brown, inner face hyaline, dark brown and concave at top; ligule 3 mm, obtuse, tubular at least on stem. *Infl.* about $\frac{1}{2}$ length of stem; bracts lf-like, at least one usually but not invariably exceeding the infl. ♂ *spike* 1, 10–30 mm; ♂ *glumes* 3–4 mm, oblong-obovate, orange-brown; apex mucronate, often fimbriate. ♀ *spikes* 2–4, upper ± contiguous, lower distant, 5–25 mm, ovoid-cylindric; peduncles ensheathed; ♀ *glumes* 2.5–3.5 mm, obovate, yellowish or pale brown with green midrib, margin hyaline; apex acuminate or obtuse and mucronate. *Utricles* 3–4 mm, obovoid-ellipsoid, ± inflated, prominently ribbed when dry, with lateral nerves prominent, shiny, pale green, minutely dotted with red-brown, inserted at an angle of 75–80° to the stem axis and therefore strongly patent, narrowing abruptly into a beak 0.75 mm, widely bifid; stigmas 3; nut obovoid, trigonous, dark brown, shortly stalked.   Fr. 7–8.

A plant of similar estuarine habitats to *C. distans*, preferring sandy soils, and on wet cliffs and raised beaches, usually within reach of the salt spray; in Britain not found inland but occasionally in non-brackish marshes with high base-status. Predominantly an oceanic species with its main stations in SW Ireland but reaching as far N as N Donegal and the Solway Firth and as far east as S Hants. Old records for the east coast are possibly *C. distans*, but a specimen from Berwick has been confirmed.

This species is frequently confused with *C. distans*, its close ally, but is usually more erect and rigid, with broader and yellower leaves. In *C. distans* none of the bracts exceeds the infl., and the ligule is not tubular; furthermore the ♂ glumes are purple-brown and the more evenly tapered utricles are not usually shiny (though they are occasionally so) and are invariably inserted more obliquely on the spike-axis. These points also differentiate *C. binervis* with which *C. punctata* has been found to form a sterile hybrid in Wales; hybrids with *C. demissa* and *C. pallescens* have been recorded from Europe.

148

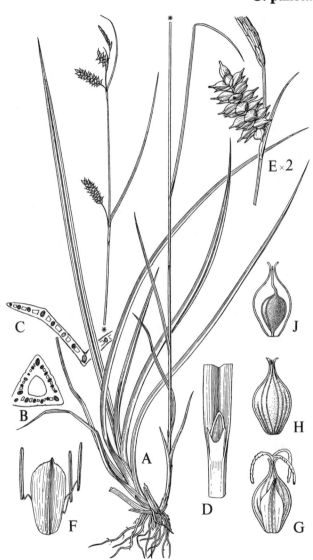

E×2

C

B

A

D

F

G

H

J

42. Carex extensa   Good.

Long-bracted Sedge   Map 41

Rhizomes short; shoots often forming large tufts; roots red-brown (often stained black); scales dark grey-brown, becoming ± fibrous. *Stems* 5–40 cm, rigid, bluntly trigonous, sometimes arcuate, solid. *Lvs* 5–35 cm × 2–3 mm, rigid, thick, ± keeled, often inrolled, gradually tapered to a blunt apex, grey-green or glaucous, becoming on dying red-brown then grey, overwintering; sheaths orange-brown, occasionally red tinged, darkening and often blackish and fibrous on decay, inner face narrow, hyaline, apex concave; ligule 2 mm, rounded. *Infl.* ⅓ or ½ length of stem; bracts lf-like, usually reflexed, far exceeding infl. ♂ *spike* usually 1, rarely 2–3, 5–25 mm; ♂ *glumes* 3–4 mm, obovate-elliptic, red-brown, with paler midrib; apex obtuse. ♀ *spikes* 2–4, contiguous or lowest sometimes distant, 5–20 mm, subglobose to cylindric; peduncles ensheathed, lowest ± exserted; ♀ *glumes* 1.5–2 mm, broadly ovate, red-brown, with pale midrib, often green, with ± hyaline margins; apex mucronate. *Utricles* 3–4 mm, ovoid or ellipsoid, weakly ribbed, grey-green or brownish with purplish blotches; beak 0.5–0.75 mm, smooth, notched; stigmas 3; nut broadly ovoid, trigonous.   Fr. 7–8.

A coastal plant, usually within reach of salt water or sea spray, on both muddy and sandy estuarine or littoral flats with *Festuca rubra*, *Puccinellia* spp., *Juncus gerardii*, *C. distans* etc. Around the coasts of Britain and Ireland, the Scottish plants usually very dwarf (forma *minor*); not recorded for Shetland.

There is no other plant in this habitat that could be easily mistaken for *C. extensa* although very small non-flowering plants on Scottish salt-marshes may be confused with forms of *C. serotina*. The exceedingly long bracts, grey-green utricles and deeply channelled, glaucous leaves are usually sufficient to identify *C. extensa*.

It forms a hybrid with *C. distans* (= *C.* × *tornabenii* Chiov.), which has been recorded for the British Isles; that with *C. demissa* is not recorded for our flora.

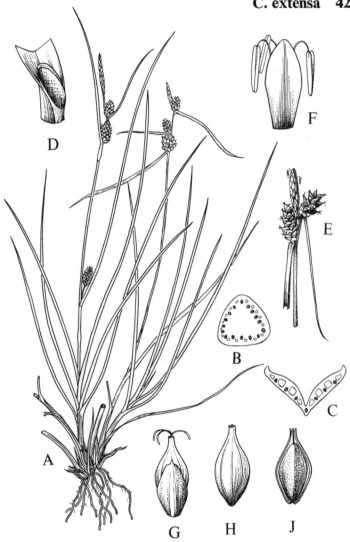

43. Carex hostiana DC.

Tawny Sedge   Map 42

Rhizome shortly creeping; shoots hardly tufted; roots of varying colours; scales pale brown, soon decaying leaving robust fibres. *Stems* 15–60 cm, slightly rough, trigonous. *Lvs* 5–30 cm × 2–5 mm, ± flat or shallowly keeled, apex abruptly contracted to a trigonous linear point (*Fig.* K), light green to yellow-green, grey-brown when dead, rarely overwintering; sheaths dark grey-brown becoming fibrous, inner sheaths pale, inner face hyaline with convex or lingulate apex; ligule 1 mm, rounded. *Infl.* ¼–½ length of stem; bracts lf-like, longer than spike but not exceeding infl., upper bract sometimes very short. ♂ *spikes* 1 or 2, 10–20 mm; ♂ *glumes* 3.5–4.5 mm, obovate-elliptic, brown with broad hyaline margin; apex obtuse. ♀ *spikes* 1–3, ± distant, 8–20 mm, ovoid-cylindric, erect; peduncles ensheathed for half their length. ♀ *glumes* 2.5–3.5 mm, broadly ovate, dark brown with broad hyaline margin and pale, often green, midrib; apex acute. *Utricle* 4–5 mm, obovoid, ribbed, yellow-green; beak 1 mm, ± deeply bifid, serrulate; stigmas 3, style often shortly exserted; nut obovoid, trigonous, shortly stalked.   Fr. 7–8.

A plant of wet flushes and marshy ground where the water contains a fairly high proportion of bases and has a pH of 5.5–6.5, e.g. common on schistose and other igneous rock flushes; often close to the shore. Common in hilly areas of N England and Scotland; more local along spring lines and in valley bogs in the S and E, and in Wales. In Ireland probably in more acid and less base-rich mires.

This species can be confused with *C. distans*; the longer beak and the contrast of the bright yellow-green utricle and dark brown ♀ glume with broad silvery margins and the trigonous point to the leaf are sufficient to separate them.

It hybridises with *C. distans* (= *C.* × *muellerana* F. W. Schultz), *C. lepidocarpa* (= *C.* × *fulva* Good.), *C. demissa* and *C. serotina* (= *C.* × *appeliana* Zahn).

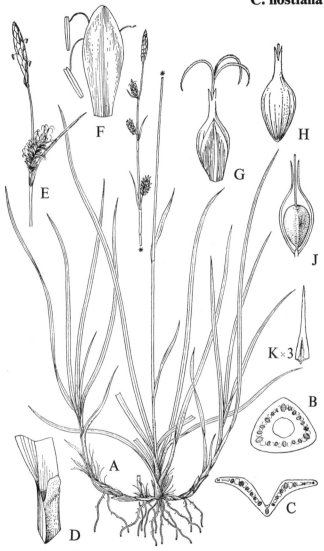

44–47. Carex flava group

44. C. flava L.
45. C. lepidocarpa Tausch
46. C. demissa Hornem.
47. C. serotina Mérat

The four taxa in this group comprise the most difficult of British sedges. The difficulties are caused by several different factors, and it is not easy to foresee any solution in terms of a readily workable taxonomy. The taxa are all extremely variable, and much of the variation seems not to be correlated either with ecological factors or with general distribution. The characters traditionally used to distinguish the taxa are largely quantitative and tend to overlap a great deal; what qualitative characters there are tend to be difficult to recognise in dried material. Considerable gene exchange is presumed to take place between several of the taxa, as intermediate and at least partially fertile populations often occur. These populations have been explained as hybrids and have in some cases been synthesized, e.g. by Davies (1956) whose views have been reiterated by Wallace (1975). Although the morphology may be clearly intermediate in some populations, in others the recombination of characters is such as to produce forms difficult to assign with certainty to either putative parent and therefore very difficult to identify. We prefer to retain the concept of hybrids for those taxa which are both morphologically intermediate and wholly sterile, a state which in this group we find only in C. demissa × flava from N Lancs. Plants intermediate between C. flava and C. lepidocarpa, on the other hand, are fertile; they occur at Malham Tarn, Lough Corrib and in N Hants, where C. flava presumably once occurred but is now extinct. Populations from Ireland difficult to assign with certainty to either C. demissa or C. serotina may well be of hybrid origin.

The relationships of the British taxa to plants elsewhere in Europe and in North America are very uncertain; for example, C. serotina appears to be conspecific with C. viridula Michx from North America and our plants should probably be called by this earlier name, though whether they comprise the same subspecies is unclear. Plants from several sites in Ireland match exactly some authentic material of C. jemtlandica (Palmgren) Palmgren from Scandinavia; they differ from C. lepidocarpa chiefly in having the leaves almost as long as the stems and the utricles more gradually narrowed into a rather straighter beak, and like *C. jemtlandica*, may be no more than a derivative of a *C. flava* and *C. lepidocarpa* cross. Other plants from various sites in Ireland match authentic material of *C. bergrothii* Palmgren, also from Scandinavia; they resemble typical *C. serotina* but have larger utricles 3–4 mm, and differ from *C. demissa* chiefly in having erect stems, narrower leaves and the utricles more abruptly narrowed into the beak. Little would therefore be gained by calling Irish plants *C. jemtlandica* or *C. bergrothii* until the variation of the group in Ireland (and indeed in the rest of its range) is more thoroughly understood. Both these "species" are perhaps best regarded as combinations of characters in the variation of the *C. flava* group which can turn up by chance more or less anywhere and there may well be no connection between Irish and Scandinavian plants with the same morphology.

Plants of *C. lepidocarpa* from Scotland and Ireland with the stems three times as long as the leaves, leaves c. 4.5 mm wide, dark green utricles and dark brown, persistent female glumes have been separated as subsp. *scotica* E. W. Davies. However, they are not always easily distinguishable from typical *C. lepidocarpa* (and furthermore often approach certain forms of *C. demissa* very closely). It therefore seems best not to regard them as subspecies.

It seems unwise at present to make any major changes, such as using the name *C. viridula*, in view of the uncertainty about the taxonomy of the whole group and the possibility of other taxonomic changes being made which would have further consequences (e.g. the sinking of *C. serotina* along with others as subspecies of *C. flava*, which would seem to be at least a partial solution to the problem).

The following hybrids exist within the group:

C. demissa × *flava* (a hybrid swarm exists in Roudsea Wood)
C. demissa × *lepidocarpa*
C. demissa × *serotina*
C. flava × *lepidocarpa* (incl. *C.* × *pieperana* P. Junge) (see above)
C. lepidocarpa × *serotina* (= *C.* × *schatzii* Kneucker)

44. Carex flava L.

Large Yellow Sedge   Map 40

Rhizomes short; shoots 2–4, tufted; roots pale buff-brown; scales bleached, becoming fibrous. *Stems* 30–70 cm, trigonous, usually solid. *Lvs* 25–70 cm × 4–7 mm, almost erect, ± flat, ± abruptly tapered to a rough tip, bright yellow-green rapidly becoming bleached to straw colour on dying, rarely overwintering; sheaths thin, ribbed, pale, becoming a bleached pink-brown and fibrous, inner face hyaline, apex ± straight; ligule 5 mm, obtuse, tubular. *Infl.* $\frac{1}{6}-\frac{1}{4}$ length of stem; bracts lf-like, often patent or reflexed, far exceeding infl. ♂ *spike* 1, 10–20 mm; ♂ *glumes* 3.5–4 mm, lanceolate-elliptic, orange-brown with pale midrib; apex subacute. ♀ *spikes* 2–4, upper often clustered, lowermost distant, 8–15 mm, ovoid, sessile except for lowest which has a partly ensheathed peduncle 4–5 cm long; ♀ *glumes* 3.5–4.5 mm, ovate-lanceolate, orange-brown, with green midrib and ± hyaline margin; apex acute. *Utricles* 6–6.5 mm, broadly elliptic, trigonous, ribbed, yellow-green to golden, patent, lower deflexed when ripe; beak 2–2.5 mm, split, deflexed; stigmas 3; nut obovoid, trigonous.   Fr. 7–8.

A rare plant in Britain and probably ousted from several former localities by hybridization. Forms intermediate between this species and *C. lepidocarpa*, usually with well formed seeds, have been found in Hampshire, Mid-West Yorks and in Ireland. In our opinion these are not *C. flava*, true material of which comes only from N Lancs, apart from an 1836 specimen from Ennerdale, Cumberland. It is a plant of base-rich fen, able to withstand some shade, often growing with *C. vesicaria*, *C. disticha* etc.

It is basically like *C. lepidocarpa* but can be distinguished by the compact infl. with the ♂ spike sometimes partly hidden by the ♀ spikes; the long bracts; the long utricle more gradually tapered into a beak 2–2.5 mm long; the longer ligule and the lvs being about as long as the stem.

Besides the hybrids mentioned on p. 155, *C. flava* hybridises in Europe with *C. binervis*, *C. distans* and *C. hostiana*. *C.* × *marshalii* Ar.Benn., described as *C. flava* × *C. saxatilis* (although *C. demissa* was most likely intended) in hb. BM is *C. lepidocarpa*.

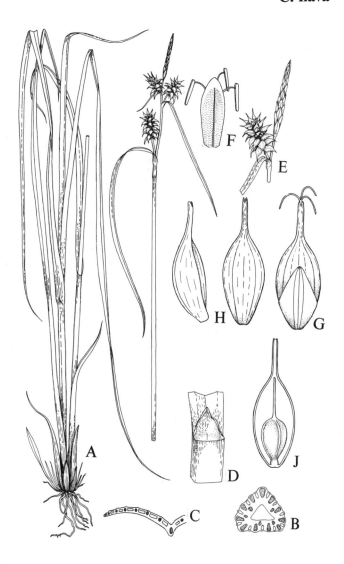

157

45. Carex lepidocarpa   Tausch

Long-stalked Yellow Sedge   Map 43

Rhizomes short; shoots 4 to several, loosely tufted; roots pale buff-brown; scales grey-brown, soon decaying. *Stems* 20–75 cm, trigonous, solid. *Lvs* 10–40 cm × 2–3.5 mm, usually only half as long as stem, keeled, ± abruptly narrowed to a rough, blunt, trigonous point, mid- to yellow-green, soon becoming bleached straw-colour on dying, not overwintering; sheaths hyaline, ribbed, soon becoming pink-brown and fibrous, apex of inner face ± straight; ligule 1 mm, rounded, tubular. *Infl.* $\frac{1}{10}$–$\frac{1}{4}$ length of stem; bracts lf-like or the upper glumaceous, occasionally reflexed, exceeding infl. ♂ *spike* 1, 10–20 mm; ♂ *glumes* 3–3.5 mm, lanceolate-elliptic, orange- or red-brown, with green midrib; apex subacute. ♀ *spikes* 2–4, upper contiguous, lower often distant, 8–15 mm, ovoid, sessile or lowermost with an ensheathed short peduncle; ♀ *glumes* falling before the utricle, 2.5–4 mm, ovate-lanceolate, orange- or red-brown, with green midrib and often hyaline margin; apex acute. *Utricles* 3.5–5 mm, obovoid-trigonous, ribbed, yellow-green, patent, lower deflexed; beak 1.5–2 mm, split, arcuate in side view; stigmas 3; nut obovoid, trigonous.   Fr. 7–8.

Predominantly a species of base-rich fens (pH 5–7.5) especially in areas with seasonal flooding or flushing and found in peaty habitats where the concentration of calcium is greater than 20 ppm; the presence of more than 1 ppm of aluminium inhibits root-growth and is thus a dominant ecological factor (Clymo 1962). Its distribution in Britain follows that of base-rich strata, sodium perhaps replacing calcium in coastal habitats.

The variation in this species grades into *C. demissa*. Usually the ratio of leaf- to stem-length is a good character to separate them. The utricle of *C. lepidocarpa* has a more gradual transition to a longer, deflexed beak; there is a distinct shoulder on the *C. demissa* utricle and the beak there is usually straight. Besides the hybrids formed within the aggregate as mentioned on p. 155, *C. lepidocarpa* hybridises with *C. distans* (= *C.* × *binderi* Podp.) and *C. hostiana* (= *C.* × *fulva* Good.). An apparent hybrid with *C. distans* has been found on the island of Danna, Kintyre. Hybrids with *C. binervis* have been recorded from Germany.

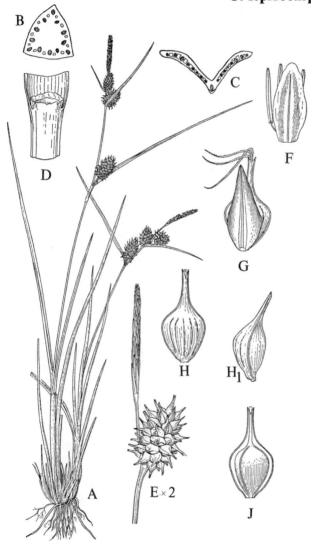

46. Carex demissa Hornem.

Common Yellow Sedge   Map 44

Rhizome short; shoots ± densely tufted; roots white to pale yellow-brown; scales grey-brown becoming fibrous and soon decaying completely. *Stems* 5–40 cm, often shorter than the leaves, subterete, solid. *Lvs* 5–35 cm × 1.5–5 mm, rigid, recurved, sharply keeled, ± abruptly tapered at apex, dark yellow-green dying to a straw colour, often overwintering; sheaths hyaline or white with green veins, persistent, becoming grey-brown, apex of inner face ± straight; ligule c. 1 mm, rounded, notched. *Infl.* in upper half but often with solitary distant spike at base of stem; bracts lf-like, often reflexed, ± flaccid and, except for the lowest, well exceeding infl. ♂ *spike* 1, 15–20 mm; ♂ *glumes* 3–4 mm, oblong-lanceolate, orange-brown to hyaline, with paler midrib; apex obtuse. ♀ *spikes* 2–4, ± contiguous but usually the lowest remote, 7–13 mm, ovoid; peduncles short, the lower up to four times as long as the spike, half ensheathed. ♀ *glumes* c. 3.5 mm, ovate, brown with green midrib; apex subacute. *Utricles* 3–4 mm, obovoid, abruptly narrowed into beak, patent, lower deflexed when ripe, yellow-green, faintly ribbed; beak 1 mm, bifid, not, or rarely slightly, deflexed (*Figs.* G[1] and H[1]) (although drying may bring about unequal collapse of the utricle apex thus deflexing the beak); stigmas 3; nut obovoid, trigonous.   Fr. 7–9.

A plant of similar habitats to *C. lepidocarpa* (i.e. with a pH of 4.5–7) but less tolerant of a high base content and usually found when the calcium concentration is less than 30 ppm, e.g. in Pennine sandstone flushes with *C. nigra*, *C. echinata* etc. It can withstand a higher aluminium concentration than *C. lepidocarpa* and is often found on inorganic soils, e.g. gravelly lake margins. Widespread in Britain and Ireland, especially common in N and W; less so in the chalk and limestone areas of E Anglia, SE Midlands and S England.

The differences between this and *C. lepidocarpa* are given on p. 158. Small stunted plants can be confused with *C. serotina* but the straight stems, often paler green leaves and the smaller, crowded utricles distinguish that species.

Besides the hybrids mentioned on p. 155, *C. demissa* hybridises with *C. binervis* (= *C.* × *corstorphinei* Druce), *C. hostiana* and *C. laevigata*; further hybrids with *C. punctata*, *C. binervis* and *C. extensa* are recorded outside the British Isles.

160

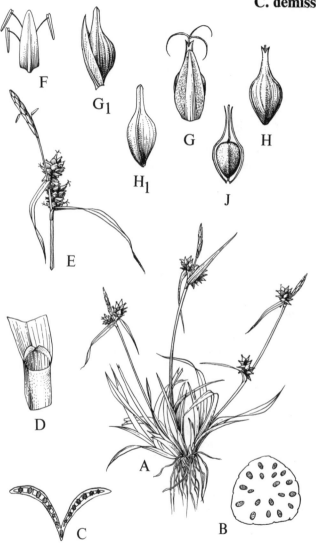

47. Carex serotina Mérat

Small-fruited Yellow Sedge Maps 45, 46

Rhizomes short; shoots 4 to many, tufted; roots pale brown; scales pink- or grey-brown, becoming fibrous. *Stems* 5–40 cm, trigonous-terete. *Lvs* 15–40 cm × 1.5–3 mm, rigid, spreading or erect, flat or ± channelled or even inrolled, gradually narrowed to a blunt tip, yellow- to grey-green, becoming pale grey-brown and persistent on dying, possibly overwintering; sheaths white or hyaline becoming grey-brown, persistent, apex of inner face straight; ligule c. 1 mm, rounded, tubular. *Infl.* up to $\frac{3}{4}$ length of stem; bracts lf-like, stiff, spreading but rarely deflexed, much exceeding infl. ♂ *spike* 1, 5–20 mm; ♂ *glumes* c. 4 mm, oblong-lanceolate, orange-brown with pale or green midrib; apex acute. ♀ *spikes* 2–5, 5–10 mm, ovoid, upper contiguous and sessile, the lower occasionally remote and pedunculate; ♀ *glumes* 2–3 mm, ovate, pale yellow-brown with green midrib; apex subacute. *Utricles* 1.75–3.5 mm, obovoid to ellipsoid, abruptly or gradually contracted into beak, faintly nerved, yellow-green, slightly inflated or not; beak 0.25–1 mm, straight or only slightly deflexed, split; stigmas 3; nut obovoid, trigonous. Fr. 7–9.

A plant of moist to wet habitats, tolerant of a wide range of pH but most frequent on acid soils in valley bogs with *C. hostiana*, *C. panicea*, *C. pulicaris* etc., in dune slacks and on stony lake shores. It is found in estuarine marshes often around the high tide level with *Puccinellia* spp., *Plantago maritima* etc; also in freshwater marshes e.g. where *Juncus acutiflorus* is dominant. Scattered throughout the British Isles mainly in lowland habitats. An inland form, known hitherto as *C. oederi* var. *cyperoides* Marsson, in which the ♂ *spike* is often replaced by a terminal ♀ one, is found in Somerset and a few other southern localities. Forms from Norfolk freshwater marshes are stoloniferous and need further investigation. Besides the hybrids mentioned on p. 155, *C. serotina* hybridises with *C. hostiana*. For confusion with *C. demissa* see p. 160.

Plants of *C. serotina*, from a wide variety of habitats, with small utricles not more than 2 mm, tightly enclosing the nut and with a very short beak (*Figs* G_2, H_2, J_2), have been separated as subsp. *pulchella* (Lönnr.) van Ooststr. (*C. scandinavica* E. W. Davies) but these plants intergrade completely with normal *C. serotina* to such an extent that any distinction is often impossible. It seems best to give this variant no formal recognition, even at subspecific rank.

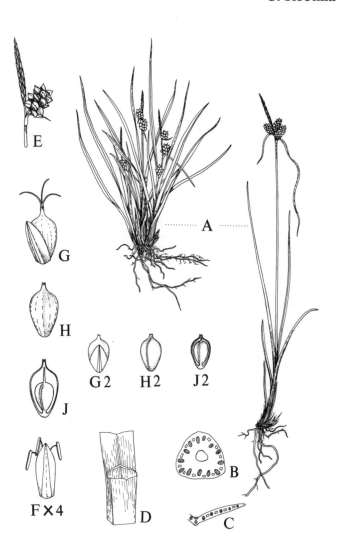

48. Carex pallescens L.

Pale Sedge   Map 37

Rhizomes very short; shoots tufted; roots usually dark red-brown; scales brown, often red-tinged. *Stems* 20–60 cm, trigonous, rough on the sharp angles. *Lvs* 15–50 cm × 2–5 mm, soft, ± hairy beneath, flat or often keeled, gradually tapering to a fine point, mid-green, grey-brown on dying; sheaths brown, hairy, persistent, inner face hyaline, hairy, apex concave; ligule c. 5 mm, acute or ± obtuse. *Infl.* up to ¼ length of stem but usually much shorter; bracts lf-like, the lower exceeding infl., crimped at base, uppermost setaceous. ♂ *spike* 1, 8–12 mm, often concealed by ♀ spikes; ♂ *glumes* 3–4 mm, obovate-oblong, pale brown, midrib often darker; apex mucronate. ♀ *spikes* 2–3, clustered or the lower remote, 5–20 mm, subglobose to ovoid, suberect or lower nodding; peduncles smooth, the lower often longer than spike; ♀ *glumes* 3–4 mm, ovate, pale brown or hyaline with a broad midrib; apex acuminate. *Utricles* c. 4 mm, ovoid-oblong, apex rounded not beaked, mid-green, shiny, faintly nerved; stigmas 3; nut ellipsoid, trigonous, stalked.   Fr. 6–7.

C. pallescens is predominantly a plant of open woodland either on heavy clays where it may form large clumps or on better drained soils where water is always available. Possibly a woodland relict in hilly places where it will grow on open wet ledges and streambanks with grasses and other sedges, although it is found in Scotland on similar wet grassy ledges at altitudes higher than forest limit. In England more frequent on the heavier soils of the SE and S Midlands and common in rough grassland of the N Pennines and SW Scotland. Common in W and Highland Scotland as a component of the acid grassland vegetation of rocky hillsides.

This species can be confused with little else. The utricle shape and the crimped base of the lower bracts are distinctive; when not fruiting the hairs on the lf-sheath and underside of leaf make it easily recognisable. Occasionally subglabrous forms are found, but hairs are always present on the inner face of the lf-sheath.

A hybrid with *C. laevigata* was discovered near Corran Ferry, v.-c. 97, by John Raven.

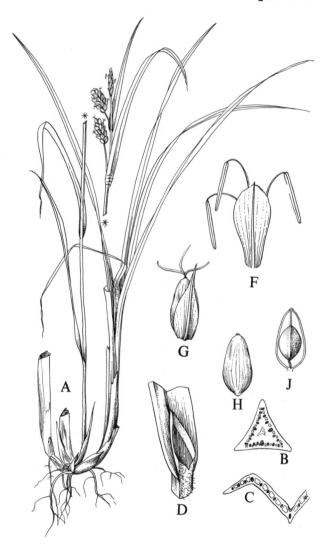

49. Carex digitata L.

Fingered Sedge   Map 48

Rhizomes short, branched; shoots tufted, of two kinds, apical lfy shoots with overwintering terminal buds and lateral short shoots ending in flowering stem; roots deep red-brown, wiry, fibrous; scales purplish-crimson. *Stems* 5–25 cm, slender, trigonous-subterete with 2–4 short (up to 2 cm), often setaceous, lvs at base. *Lvs* 5–25 cm × 1.5–5 mm, usually sparsely hairy on upper surface, soft, ± flat or slightly keeled, tapered abruptly to a blunt point, light to bronzy-green, often overwintering; margins rough; sheaths bright crimson (even in very young shoots), inner face herbaceous, apex concave soon splitting; ligule 0.5–1.5 mm, rounded. *Infl.* $\frac{1}{5}$ to $\frac{1}{4}$ length of stem, the lowest branch distant 1 cm or more from the next above; bracts glumaceous. ♂ *glumes* 5 mm, red-brown with a pale midrib and hyaline margin, rounded or even emarginate at apex. ♀ *spikes* 1–2 cm, 5–10-fld, pedunculate; ♀ *glumes* 3–4.5 mm, obovate, purplish-crimson; apex obtuse or emarginate. *Utricles* 3–4.5 mm, obovoid, greenish brown, pilose; beak almost 0, truncate; stigmas 3; nut obovoid, trigonous, stalked.   Fr. 4–6.

In open woodland or scrub and amongst rocks and on stabilised screes of hard chalk and limestone. Flowers earlier than No. 50, often in early April, and can withstand more shaded conditions when growing occasionally in open woodland or scrub. Very local but more widely spread than *C. ornithopoda*, extending from Somerset and Dorset to Westmorland and NE Yorks.

The purplish glumes as long as the utricles and the separation of the branches of the inflorescence distinguish *C. digitata* from *C. ornithopoda*, in which the straw-coloured glumes are shorter than the utricles and the branches all spring from nearly the same point. Vegetatively *C. digitata* is usually larger in all its parts and the sheaths of even the youngest shoots are deeply tinged with crimson.

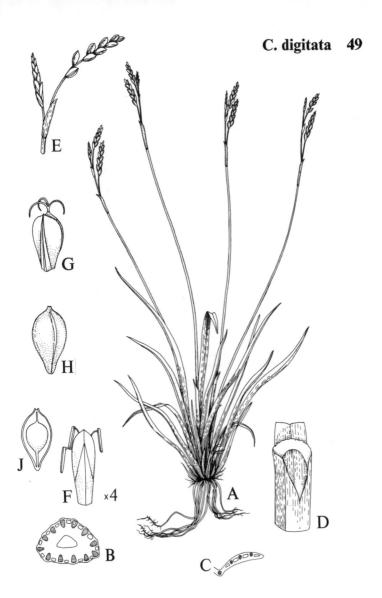

50. Carex ornithopoda Willd.

Bird's-foot Sedge   Map 47

Rhizomes short, much branched; shoots tufted, of two kinds, apical lfy shoots with overwintering terminal buds and lateral short shoots ending in flowering stem; roots deep brown, wiry, fibrous; scales deep red-brown, becoming fibrous. *Stems* 5–20 cm, slender, trigonous-subterete with 2–4, short (up to 1 cm), often setaceous, lvs at base. *Lvs* 5–20 cm × 1–3 mm, soft, ± flat or slightly keeled, tapered abruptly to a blunt point, light or mid green, often over-wintering; margins rough; sheaths orange- to dark crimson-brown, becoming fibrous, inner face herbaceous, apex concave soon splitting; ligule 0.5–1 mm, rounded. *Infl.* $\frac{1}{8}$–$\frac{1}{10}$ length of stem, compact, ± digitate; bracts glumaceous. ♂ *spike* 1, 5–8 mm, few-fld, overtopped by the lower ♀ spikes and appearing lateral; ♂ *glumes* c. 2.5 mm, obovate, red-brown, with a pale midrib and hyaline margin; apex acute or mucronate. ♀ *spikes* 2–3, 5–10 mm, 2–4 fld, ± sessile; ♀ *glumes* 2–2.5 mm, obovate, pale orange-brown with hyaline margin; apex obtuse or ± acute, often erose. *Utricles* 2–3 mm, obovoid, pyriform, yellow-green to brown, pilose; beak almost 0, truncate; stigmas 3; nut obovoid, trigonous, stalked.   Fr. 5–6.

A plant of dry, well-drained limestone grassland or in crevices in limestone pavement, rarely in the shade. Very local: in Derby, Yorkshire and Westmorland. Reported from Cumberland but the report is not confirmed.

Closely related to and sometimes confused with *C. digitata* with which it hybridises in Europe.

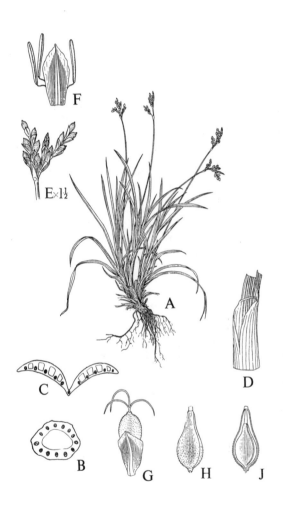

51. Carex humilis Leysser

Dwarf Sedge

Map 47

Rhizomes shortly creeping, often much branched; shoots densely fasciculate at branch tips; roots purple-brown, woody, much branched; scales red-brown, persisting as fibres. *Stems* 2–10 (rarely 15) cm, slender, often arcuate or flexuous, subterete, solid, often hidden by lvs. *Lvs* up to 20 cm × 1–1.5 mm, rough, stiff, arcuate, at first flat, later becoming channelled, tapering from base to a fine trigonous point, dark green, pale purple-brown on decay, over-wintering; sheaths white with green veins, becoming orange- and red-brown, persistent, eventually becoming fibrous and clothing rhizome, inner face hyaline, apex concave; ligule 0.5–1 mm, rounded. *Infl.* up to ¾ length of stem; bracts glumaceous, hyaline or pale brown, almost enclosing ♀ spike and with a sheath 3–8 mm. ♂ *spike* 1, 10–15 mm; ♂ *glumes* 5–7 mm, elliptic-oblanceolate, red- or purple-brown with very broad hyaline margin and pale midrib; apex obtuse or subacute. ♀ *spikes* 2–4, distant, 4–10 mm, fusiform, with 2–4 fls only; peduncles 1–2 mm, ensheathed; ♀ *glumes* 2–3 mm, obovate to broadly elliptic, clasping utricle and appearing narrower, red-brown, hyaline at edges and base; apex obtuse, often mucronate. *Utricle* c. 2.5 mm, obovoid, pyriform, trigonous, shortly hairy; beak almost 0, truncate; stigmas 3; nut ± ellipsoid, trigonous, stalked.   Fr. 4–7.

C. humilis is a species of short turf in species-rich limestone and chalk grassland and usually but not exclusively found on S, SW and W facing slopes in such associations. Locally abundant: most frequent on the Dorset and S Wiltshire downs; also in S Hampshire, N Wilts, N Somerset, W Gloucester and Hereford.

There is no other *Carex* species that *C. humilis* could be confused with in calcareous grassland except perhaps depauperate non-fruiting *C. montana*, the soft leaves of which are in contrast to the rigid, arcuate leaves of *C. humilis*. The latter is easily overlooked because of its narrow, *Festuca*-like leaves; the infl. must be searched for well within the tuft in early spring—both glumes and utricles drop towards the end of the season.

No hybrids are known.

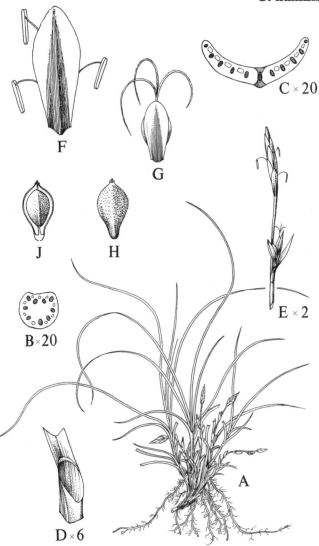

52. Carex caryophyllea   Latourr.

Spring Sedge   Map 51

Rhizomes shortly creeping; shoots loosely tufted; roots dark- or purple-brown; scales rich or black-brown, often shiny, soon becoming fibrous. *Stems* 2–30 cm, trigonous, leafy below. *Lvs* up to 20 cm × 1.5–2.5 mm, recurved, rough on upper surface, shiny, ± flat, tapered ± abruptly to a short trigonous point, mid or dark green, sometimes overwintering; sheaths herbaceous, becoming brown and fibrous, inner face hyaline, soon decaying, apex straight; ligule 1–2 mm, obtuse, free margin very narrow. *Infl.* 2–4 cm; lower bracts often lf-like, the lowest with a sheath 3–5 mm, the upper glumaceous. ♂ *spike* 1, 10–15 mm, clavate; ♂ *glumes* 4–5 mm, oblanceolate-elliptic, red-brown, hyaline towards base, with darker midrib; apex ± acute or mucronate. ♀ *spikes* 1–3, ± contiguous, 5–12 mm, ovoid, erect, sessile or with peduncles ensheathed; ♀ *glumes* 2–2.5 mm, broadly ovate, red-brown, with excurrent green midrib; apex ± obtuse, often attenuate. *Utricle* 2–3 mm, obovoid-ellipsoid, ± trigonous, with two lateral ribs, tomentose, green; beak c. 0.2 mm, notched; stigmas 3; nut obovoid-ellipsoid trigonous.   Fr. 5–7.

C. caryophyllea is a plant of wide ecological tolerance. Frequent in dry calcareous grassland in S and E England and on acid soils on mountains where flushed occasionally by base-rich water. Throughout the British Isles but local N of the Great Glen.

Similar to *C. ericetorum* (q.v.); it can be distinguished from small *C. binervis* by its blackish-brown fibrous sheaths and flatter, shiny leaf; from *C. pilulifera* see p. 180. *C. caryophyllea* hybridises with *C. pilulifera* and *C. ericetorum* but neither hybrid is so far reported from Britain.

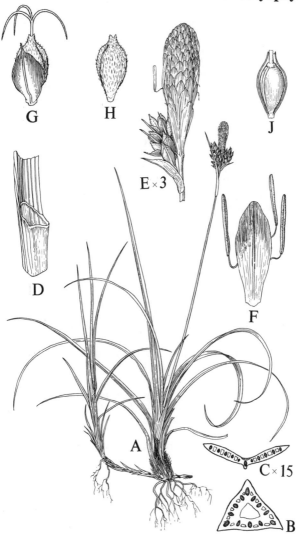

G

H

E × 3

J

D

F

A

C × 15

B

53. Carex tomentosa L.

Downy-fruited Sedge   Map 49

Rhizomes long creeping, often very slender; shoots 2–3 per tuft; roots pale; scales red-brown, shiny, with sharp points, persistent. *Stems* 20–50 cm, rough above, trigonous, slender. *Lvs* 15–40 cm × 1.5–2 mm, rough, flat, gradually tapering to a fine point, ± glaucous, grey-brown on decay, often overwintering; sheaths red or red-purple, inner face hyaline, becoming brown, often persistent and fibrillose on splitting, apex concave; ligule 1–2 mm, obtuse, tubular. *Infl.* $\frac{1}{6}$ or less of stem length; upper bracts setaceous, lower lf-like, ± equalling infl., the lowest not or shortly sheathing. ♂ *spikes* 1–2, 12–25 mm; ♂ *glumes* 4–5 mm, elliptic-ovate, red-brown with paler midrib and ± hyaline margin; apex ± acute, often apiculate. ♀ *spikes* 1–2, ± contiguous, 5–14 mm, oblong-ovoid to subglobose; ♀ *glumes* 2–3 mm, ovate to subrotund, purple- or red-brown with pale midrib; apex acute or lowermost in spike mucronate. *Utricles* 2–3 mm, subglobose or pyriform, trigonous, tomentose, green; beak 0.1–0.2 mm, notched; stigmas 3; nut obovoid or pyriform, trigonous.   Fr. 6–7.

C. tomentosa is a local plant of fairly rich pastures (Timothy/Rye Grass association) and of damper meadows, roadsides and rough ground in conjunction with other sedges e.g. *C. panicea*, *C. flacca*, *C. nigra* etc. Recorded for Surrey, Middlesex, E Gloucester and Oxford and N Wilts. Its limited distribution in Britain is not easy to explain.

The small hairy utricles of this species distinguish it from any other British sedge of comparable size; the lf-like bracts of *C. tomentosa* separate it from *C. caryophyllea*. Vegetatively it may be similar to *C. flacca* and other glaucous sedges of its habitat (e.g. *C. nigra*) but the shiny, red basal lf-sheaths and scales, and the brown, persistent inner face of the lf-sheath which usually shows some fibrillae, serve to distinguish it.

A hybrid with *C. flacca* has been described from France.

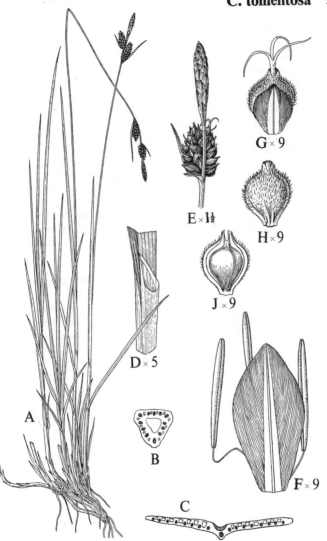

A

B

C

D × 5

E × 1½

F × 9

G × 9

H × 9

J × 9

54. Carex ericetorum Pollich

Rare Spring Sedge   Map 52

Rhizomes shortly creeping; shoots tufted, forming a close mat; roots dark- or purple-brown; scales rich or black-brown, becoming fibrous. *Stems* 2–20 cm, bluntly trigonous, ± leafless or with 3 very short lvs at base. *Lvs* up to 15 cm × 1.5–4 mm, often recurved, rough sometimes almost papillose on upper surface, shiny, ± flat, tapered ± abruptly to a short trigonous point, mid or dark green with usually a broad scarious margin, sometimes overwintering; sheaths herbaceous, becoming brown and fibrous; ligule less than 1 mm, rounded. *Infl.* 2–3 cm, bracts glumaceous, the lowest scarcely sheathing. ♂ *spike* 1, 10–15 mm, narrowly cylindrical; ♂ *glumes* 2–3 mm, oblong, purple-brown, margin scarious, fringed; apex rounded. ♀ *spikes* 1–3, ± contiguous, 5–12 mm, ovoid, erect, sessile; ♀ *glumes* 2–2.5 mm, rounded, purple-brown becoming black, margin scarious, fringed. *Utricle* 2–3 mm, subglobose-trigonous, tomentose, dark green, beak very short; stigmas 3; nut subglobose.
Fr. 4–6.

In dry calcareous grassland with *C. caryophyllea*, from which it is not easy to separate by vegetative characters. When in flower or fruit the dark purplish glumes, which in the ♀ spike are obtuse with a broad scarious and often fimbriate margin, provide an immediate identification. Local, on the E Anglian chalk, and on the limestone in Lincs, Derby, Yorks, Durham, and Westmorland. Hybrids with *C. caryophyllea*, *C. pilulifera*, and *C. montana* are reported, but not from Britain.

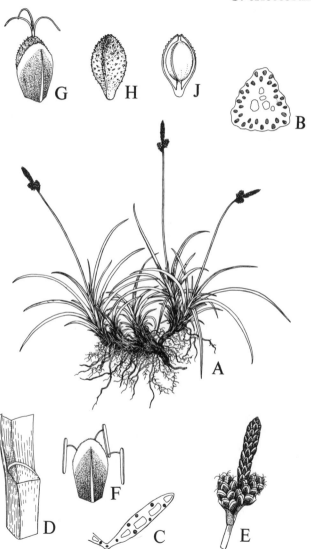

55. Carex montana L.

Soft-leaved Sedge   Map 53

Rhizomes creeping, thick, often much branched; shoots tufted at apex; roots purple-brown, woody; scales red-brown, becoming fibrous. *Stems* 10–40 cm, slender, rough at top, flaccid, trigonous or often with 6 angles, ± solid, leafless or with few short lvs at base. *Lvs* 10–35 cm × 1.5–2 mm, soft, flat, gradually drawn out to a slender point, sparsely hairy on upper surface, becoming glabrous, light to mid green, grey-brown and persisting on dying; sheaths dark red-brown to almost bright red, ribbed, becoming fibrous and densely clothing rhizome, inner face hyaline, soon decaying, apex concave; ligule c. 1 mm, obtuse. *Infl.* 1–2 cm, very congested; bracts glumaceous or lowest setaceous, the lowest scarcely sheathing. ♂ *spike* 1, 10–20 mm; ♂ *glumes* 4–5, oblanceolate or broadly elliptic, red-brown, with pale midrib; apex acute or ± mucronate. ♀ *spikes* 1–4, 6–10 mm, ovoid, with a few lax fls, clustered beneath ♂ spike, erect, sessile; ♀ *glumes* 3–5 mm, broadly ovate or obovate, reddish-black with pale midrib and hyaline margin; apex obtuse or even retuse, mucronate. *Utricles* 3.5–4 mm, obovoid-pyriform, bluntly trigonous, tapered to a stout stalk, densely hairy, lightly ribbed, brown or blackish on exposed face; beak almost 0, notched; stigmas 3; nut obovoid, trigonous, stalked.   Fr. 5–6.

C. montana is a plant of rough grassy situations, usually on mature soils with a high base content, derived either from limestone or mineral-rich igneous rocks. Associated mainly with limestone areas and a badly named sedge as far as its British ecological range is concerned. Local, scattered across S England and Wales as far north as Derbyshire; not recorded for Ireland.

The soft mid to pale green leaves and mat-forming rhizomes (which may die out in the middle leaving a ring of sedge tufts) distinguish this species from the closely related, single-tufted, stiff-leaved, *C. pilulifera*. In flower or fruit, if the slender stems lie amongst the leaves, it may resemble *C. pilulifera*, but the mucronate ♀ glume and large pyriform utricles of *C. montana* are sufficient to identify it with certainty. *C. montana* is reported to hybridise with *C. ericetorum*, *C. flacca* and *C. pilulifera* but these hybrids are not recorded for Britain.

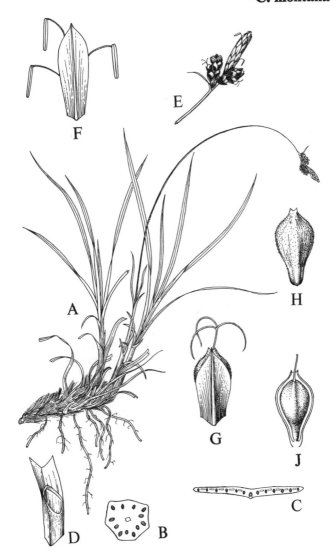

56. Carex pilulifera L.

Pill Sedge   Map 54

Rhizomes short; shoots densely tufted, often decumbent, roots purple-brown; scales brown, soon becoming fibrous. *Stems* 10–30 cm, ± rough above, wiry, sharply trigonous, often arcuate. *Lvs* 5–20 cm × 1.5–2 mm, rough above, ± flat, papillose on upper surface, ± abruptly tapered to a short trigonous point, yellow- or mid-green, pink-brown and persistent on decay, overwintering; sheaths pale red-brown or wine-red, becoming fibrous, inner face hyaline, apex straight; ligule 0.5–1 mm, rounded. *Infl.* 2–4 cm, clustered at apex of stem; bracts lf-like or upper setaceous, rarely exceeding infl., the lowest not sheathing. ♂ *spike* 1, 8–15 mm; ♂ *glumes* 3.5–4 mm, oblanceolate-elliptic, brown or chestnut, hyaline towards margin, with pale midrib; apex acute. ♀ *spikes* 2–4, 5–8 mm, ovoid or subglobose, erect, sessile; ♀ *glumes* 3–3.5 mm, broadly ovate, red-brown, hyaline towards margin, with green midrib; apex acute or acuminate. *Utricle* 2–3.5 mm, obovoid-ellipsoid, ± downy, green; beak 0.3–0.5 mm, notched; stigmas 3; nut obovoid, trigonous.

Fr. 6–7.

C. pilulifera is a plant of leached, skeletal, sandy or peaty soils or of more loamy soils with a low base content with a pH range usually between 4.5 and 6.0. A frequent plant on much of our *Agrostis-Festuca-Anthoxanthum-Nardus* hill pasture. Also in open *Calluna-Erica* heath, often taking advantage of heath-fires to re-establish and spread. Throughout Britain more frequent in sandy regions and common in Scotland where it ascends to 750 m (2500 ft); scattered, but less frequent, in Ireland.

Distinguished from *C. ericetorum* and *C. caryophyllea* by its tufted habit, reddish sheaths, dull arcuate lvs and wiry often incurved stems; the closely related *C. montana*, itself tufted and with wiry stems, differs in its more obvious creeping rhizome, softer leaves and purple-brown glumes. Further, *C. pilulifera* is a plant of more acid soils than either of the other species although local pockets of soil may bring them close together. *C. pilulifera* varies in length of bracts and ♀ glumes; one extreme form with distant spikes and bracts exceeding the infl. has been referred to in literature as forma *longe-bracteata* Lange.

C. pilulifera is reported to hybridise with *C. caryophyllea*, *C. ericetorum* and *C. montana* but these hybrids have not been recorded for the British Isles.

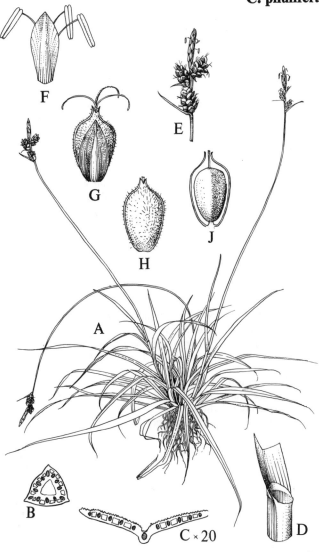

57. Carex atrofusca Schkuhr

Scorched Alpine Sedge Map 48

Rhizomes shortly creeping; shoots in loose tufts; roots yellowish; scales pale brown, soon decaying. *Stems* 5–35 cm, trigonous. *Lvs* 2–12 cm × 2–5 mm (those of sterile shoots often less than 1 mm), soft, ± flat, or slightly keeled; stem-lvs abruptly tapered to a short trigonous point, those of sterile shoots more gradually tapered to a finer trigonous point 1–3 cm long, mid green, very rarely glaucous; sheaths pale with green veins, becoming pale yellow-brown, soon decaying, inner face hyaline, apex concave; ligule 2–2.5 mm, obtuse. *Infl.* $\frac{1}{6}$–$\frac{1}{4}$ length of stem; lower bracts narrowly lf-like, shorter than the spike, the lowest with a sheath 5–15 (–25) mm, upper glumaceous. ♂ *spike* 1, 5–10 mm, broadly ellipsoid; ♂ *glumes* 3.5 mm, ovate-elliptic, dark red-brown, rarely pale, with pale or green midrib; apex acute to acuminate. ♀ *spikes* 2–4, clustered, lowest rarely distant, 5–12 mm, ovoid-globose, nodding; peduncles smooth, up to 3 times as long as spike, half-ensheathed; ♀ *glumes* c. 3 mm, oblanceolate, purple- or red-black with pale, thin or often indistinct midrib; apex acuminate. *Utricles* 4–4.5 mm, narrowly obovoid or ellipsoid, purple-black; beak 0.3 mm, notched; stigmas 3; nut ± ellipsoid, trigonous, stalked. Fr. 7–9.

A rare plant of micaceous stony flushes between 540 m (1800 ft) and 1050 m (3500 ft). With *C. demissa, C. panicea, Juncus triglumis, Pinguicula vulgaris, Thalictrum alpinum, Blindia acuta*, etc. In mid-Perth and Westerness and Argyll; the record for Rhum is in error.

It could be confused with *C. atrata* but the position of the ♂ florets and the softer less glaucous leaves and lack of brown scales is sufficient to distinguish them.

No hybrids of *C. atrofusca* are recorded.

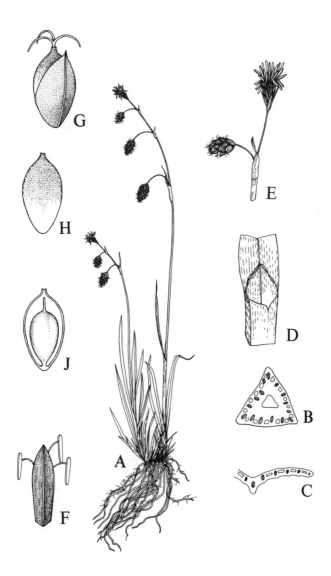

58. Carex limosa L.

Mud Sedge   Map 50

Rhizomes often far creeping, partly ascending; shoots loosely tufted, initially decumbent, slender; roots yellow, felted; scales brown, persistent. *Stems* 10–40 cm, rough, slender, rigid, trigonous, striate, often decumbent at base. *Lvs* 15–40 cm × 1–2 mm, ± rough, thin, keeled, gradually tapered to a fine rough point, pale to bluish-green, often rich brown on decay; sheaths red-flushed, becoming brown or red-brown, persistent, inner face hyaline, persistent, apex concave; ligule 3–4 mm, obtuse. *Infl.* c. ⅙ length of stem; bracts lf-like, lowest about as long as spike, not sheathing or with a sheath less than 5 mm. ♂ *spike* 1, 10–25 mm; ♂ *glumes* 3–4 mm, broadly lanceolate, red-brown, often with a green midrib; apex ± acute, apiculate. ♀ *spikes* 1–3, 7–20 mm, ± ovoid, with up to 20 fls, usually nodding; peduncles up to 40 mm, slender, smooth; ♀ *glumes* 3.5–4.5 mm, slightly wider than utricle, ovate, brown or red-purple with a green midrib; apex acute. *Utricles* 3–3.5 mm, obovoid to broadly ellipsoid, compressed, strongly ribbed, blue-green; beak 0.5 mm, truncate; stigmas 3; nut obovoid, trigonous.   Fr. 6–9.

C. limosa grows at edges of pools or in very wet blanket or marginal valley mire, with *Sphagnum subsecundum*, *S. cuspidatum*, *Menyanthes trifoliata*, *Eriophorum angustifolium*, *Utricularia minor* etc; also in mesotrophic mires with *C. nigra*, *C. lasiocarpa*, *Phragmites* etc. Usually found below 450 m (1500 ft) alt. but ascending to 817 m (2725 ft) in Breadalbane, and mainly occurring in N Wales, N England and Scotland where it becomes frequent in the blanket-mires of the west; in Ireland mainly in the N and W. Also in the S Dorset-Hampshire valley mires and in E Norfolk where it is decreasing due to drainage.

Similar in appearance and habitat to *C. magellanica*, but that species forms more solid clumps, with at least some of the lvs 2 mm or more wide, smooth except at the tip, and of a clearer green (not glaucous); the lowest bract usually overtops the inflorescence; the lowest spike has ♂ flowers at the base; and, furthermore, the ligule is longer and has a wider free margin than in *C. limosa*. When in fruit the two are easy to separate, for *C. magellanica* has acuminate ♀ glumes, narrower than the utricle. The hybrids with *C. magellanica* and *C. rariflora* are recorded in Europe.

184

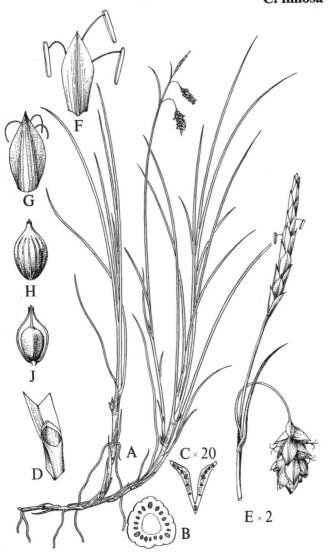

59. **Carex rariflora** (Wahlenb.) Sm.

Loose-flowered Alpine Sedge   Map 47

Rhizomes shortly creeping, often producing a close carpet of single shoots; roots yellow, felted; scales red-brown. *Stems* up to 20 cm, trigonous, $\pm$ solid. *Lvs* up to 15 cm $\times$ 1–2 mm, with c. 9 veins, often strongly incurved at the tips; sheaths becoming fibrous. *Infl.* $\frac{1}{3}$ to $\frac{1}{5}$ length of stem; lower bracts narrow, shorter than infl. ♂ *spike* 1, 8–12 mm; ♂ *glumes* c. 4 mm, oblong-ovate, dark brown; apex obtuse, mucronate. ♀ *spikes* usually 2, rarely 3, with up to 8 fls; ♀ *glumes* 3–4 mm, dark purple or chocolate-brown, oblong-ovate; apex obtuse, mucronate. *Utricles* 3–4.5 mm, ellipsoid, narrower and shorter than ♀ glumes, strongly ribbed, markedly tapered at either end; stigmas 3; nut stalked, ellipsoid, trigonous.

Fr. 7–8.

A rare and local sedge forming a characteristic association with *C. aquatilis*, *C. curta* and *Sphagnum* spp. on wet slopes of oligotrophic peat at between 750 and 1050 m (2500 and 3500 ft) in E Central Highlands; also occurring very rarely in Breadalbane. A record from S Uist of this sedge, otherwise confined to high tablelands where snow lies late, must be erroneous. It is very shy-flowering in some seasons, and this may have concealed its presence, and certainly its abundance, in a number of localities. In Britain it is separated from its two allies *C. limosa* and *C. magellanica* as clearly by the different altitudes that they normally inhabit as by its form and much smaller size. *C. rariflora* hybridises with both these species but neither hybrid has been recorded from Britain.

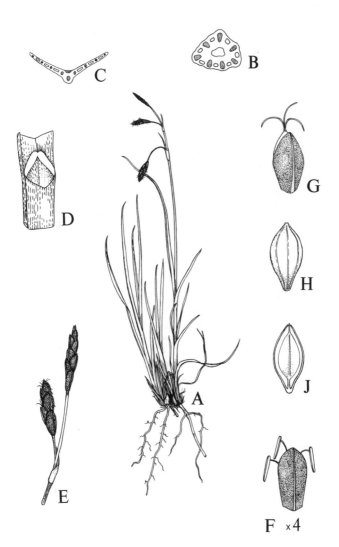

60. Carex magellanica Lam.

Bog Sedge Map 49

Rhizomes shortly creeping, often ascending; shoots loosely tufted; roots yellow, felted; scales red-brown, persistent, shiny. *Stems* 12–40 cm, trigonous, ± solid. *Lvs* up to 25 cm × 1.5–4 mm, soft, thin, with c. 15 veins, rough above, ± flat, abruptly tapered to a rough tip, pale or apple-green; sheaths becoming red-brown, persistent, inner face hyaline, apex concave; ligule c. 5 mm, acute, tubular. *Infl.* c. $\frac{1}{5}$ length of stem; bracts lf-like, lower usually exceeding infl., the lowest not or scarcely sheathing. ♂ *spike* 1, 10–20 mm; ♂ *glumes* 5–6 mm, lanceolate-elliptic, pale red-brown, hyaline at margins; apex acute. ♀ *spikes* 2–4, 5–18 mm, ovoid, lax and up to 10-fld, upper ± erect, lower nodding; peduncles smooth, 5–20 mm; ♀ *glumes* 5–6.5 mm, narrower than the utricle, lanceolate, caducous, red- or purple-brown with a paler midrib; apex acuminate or aristate. Utricles 3–3.5 mm, ovoid-globose, ± compressed, faintly ribbed, blue-green; beak 0; stigmas 3; nut ellipsoid, trigonous.

Fr. 6–7.

C. magellanica is a plant of *Sphagnum* bogs but, unlike *C. limosa*, cannot tolerate standing water. Its favourite habitat is a level ledge, near the crest of a moor, wet, but being close to the watershed, neither inundated nor subject to much movement of the water; it is for this reason that it is frequently found on county boundaries, which often run along a watershed. *C. magellanica* is rarer than *C. limosa*, but may have been overlooked or confused with that species (for differences see p. 184). Scattered from the Lake District to N Uist and W Sutherland; very local in Antrim and North Wales. Hybrids between this species and *C. rariflora* are recorded for Sweden but not for Britain.

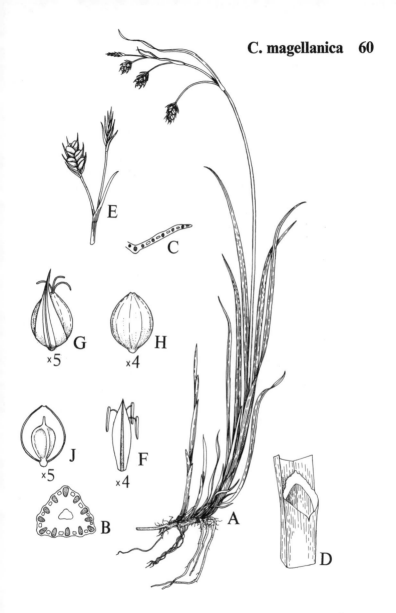

61. Carex atrata L.

Black Alpine Sedge Map 55

Rhizomes short; shoots tufted; roots pale brown; scales brown- or red-purple, persistent. *Stems* 30–55 cm, thick, rigid, trigonous, solid. *Lvs* 15–30 cm × 2–6 mm, ± flat or slightly keeled, abruptly tapered to a rough apex, glaucous; sheaths dark brown, often wine-red, persistent, inner face hyaline-brown, persistent, apex straight; ligule 2–3 mm, obtuse. *Infl.* $\frac{1}{8}-\frac{1}{6}$ length of stem; bracts lf-like, lower exceeding infl., the lowest not or scarcely sheathing. *Spikes* 3–5, ovoid, 8–20 mm, clustered in a ± nodding head, terminal ♀ at top, ♂ at base, lower ♀ only; lower peduncles rough, as long as spike. ♂ *glumes* 4.5–6 mm, ovate-elliptic, red-black with pale midrib; apex acute or rounded. ♀ *glumes* 3.5–4.5 mm, ovate-elliptic, purple or red-black with pale midrib; apex ± acute *Utricles* 3–4 mm, obovoid-ellipsoid, compressed at edges, green; beak 0.3–0.5 mm, slightly notched; stigmas 3; nut obovoid, trigonous.   Fr. 7–9.

   C. atrata is found on wet ledges in Scotland, Wales and northern England above 720 m (2400 ft) altitude usually where there is calcareous veining, often with *C. bigelowii* and occasionally with *C. atrofusca*, with the latter of which it can be confused. The lack of a terminal ♂ spike and the bract exceeding the infl. will distinguish it when in flower. When sterile, the lack of a creeping rhizome is sufficient although that of *C. atrofusca* may not be obvious; the latter has pale, often yellow-brown leaf-sheaths and rhizome scales.

   Hybrids between *C. atrata* and *C. norvegica* are known but are not recorded for Britain.

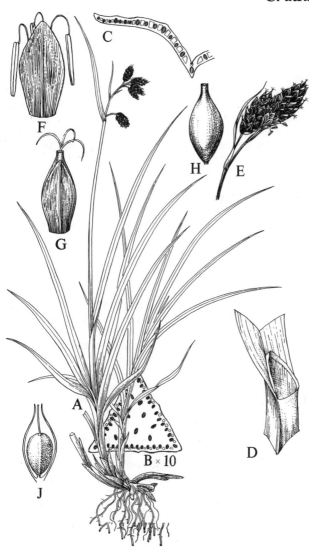

62. Carex buxbaumii Wahlenb.

Club Sedge Map 52

Rhizomes shortly creeping; shoots single or in small tufts; roots pale yellow-brown, much branched; scales red-brown, or sometimes blackish, shiny, persistent. *Stems* 30–70 cm, rigid, sharply trigonous. *Lvs* 25–60 cm × 1.5 2 mm, ± flat or keeled, gradually tapered to a fine rough point, glaucous; sheaths red- or orange-brown, persistent, inner face hyaline-brown, fibrillose on splitting; ligule c. 3 mm, acute. *Infl.* c. ¼ length of stem; upper bracts glumaceous or setaceous, lowest lf-like, ± equalling infl, not or scarcely sheathing. *Spikes* 2–5, contiguous or ± remote, 7–15 mm, erect, the terminal ♀ at top, ♂ at base, the lower spikes ♀; peduncles 2–12 mm, rough. ♂ *glumes* 4 mm, lanceolate-elliptic, red-black with pale midrib: apex acute. ♀ *glumes* 3–5 mm, dark red-brown with pale midrib; apex acuminate or aristate. *Utricles* 3–4.5 mm, indistinctly ribbed, pale green, oblong-ovoid, ± inflated, apex conical, truncate; beak 0; stigmas 3; nut ellipsoid, trigonous. Fr. 7–8.

C. buxbaumii is a rare sedge of mesotrophic fen with *Phragmites*, *Eriophorum angustifolium*, *C. lasiocarpa*, *C. nigra*, *C. rostrata* etc. Known from two localities in Scotland but possibly overlooked elsewhere; first discovered in British Isles in 1835 around L. Neagh (Antrim), where it has disappeared since draining. It is a species that dies back completely by the end of August.

Superficially like *C. nigra* but the red fibrillose leaf-sheaths, three stigmas and the terminal ♀ florets are sufficient to distinguish it (although very occasionally *C. nigra* has ♀ fls at the top of the ♂ spike). Distinguished from *C. lasiocarpa* by the acute ligule.

C. hartmanii Cajander, a plant of C Europe and Fennoscandia, has been identified from amongst the many gatherings of *C. buxbaumii* from L. Neagh; the identity of these specimens needs further investigation. *C. hartmanii* has broadly obovoid to sub-globose utricles which are 2–3 mm long and equal in length to the glumes. A northern taxon, *C. buxbaumii* subsp. *alpina* (Hartman) Liro, which has similar utricles to *C. hartmanii* but shorter, acute but not acuminate, ♀ glumes, could occur in alpine Scotland. It also has all spikes cylindrical whilst in *C. buxbaumii* the terminal spike is clavate and the others ovoid or obovoid.

No hybrids between *C. buxbaumii* and other species in this section have been recorded; those recorded with *C. acutiformis* and *C. nigra* are probably aberrant forms of these species.

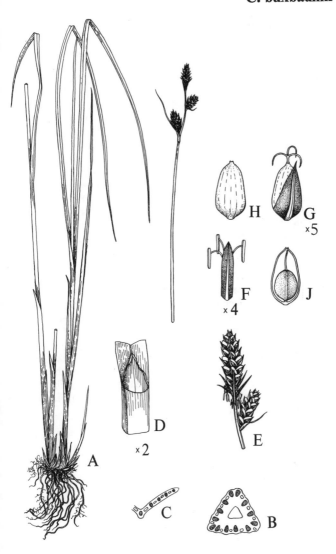

63. Carex norvegica Retz.

Close-headed Alpine Sedge Map 53

Rhizomes short; shoots tufted; roots pale yellow-brown: scales brown, red-brown or blackish, becoming fibrous on decay. *Stems* 6–30 cm, trigonous. *Lvs* 5–20 cm × 1.5–3 mm, $\pm$ flat or keeled, gradually tapered to a short, fine trigonous point, mid green, not glaucous; sheaths white, green-veined, becoming red, red-brown or brown, persistent, inner face hyaline, apex straight; ligule 0.5–1 mm, obtuse. *Infl.* a compact, terminal cluster of 1–4 subglobose, erect *spikes* the terminal one of which is ♀ at the top and ♂ at the base (often only to be detected by the presence of white filaments), the rest being ♀; upper bracts glumaceous, lower bracts lf-like, exceeding infl., the lowest not or scarcely sheathing. ♂ *glumes* 2–2.5 mm, lanceolate-elliptic, dark red-brown with pale midrib; apex acute. ♀ *glumes* 1.5–2 mm, ovate, red-black with narrow, pale midrib; apex acute. *Utricles* 2–2.5 mm, obovoid-ellipsoid, minutely papillose, greenish-brown; beak 0.25 mm, slightly notched; stigmas 3; nut ellipsoid, trigonous. Fr. 7–8.

A very rare plant of wet ledges and rocky slopes, with other sedges, grasses and dwarf-heath species. On mountains from 690–990 m (2300–3300 ft) in N facing corries; in Perth, Angus and S Aberdeen. Records from Rhum and N and S Uist must be discounted for this is a plant of the higher mainland mountains where snow lies late.

Sometimes confused with *C. bigelowii* (or *C. nigra*) but those species never form such dense tufts, have far-creeping rhizomes, two stigmas and their terminal spike is entirely ♂ (although aberrant forms of these species may have ♀ fls at the top of the ♂ spike).

C. norvegica hybridises in Sweden with *C. atrata*; this hybrid has not been reported from Britain.

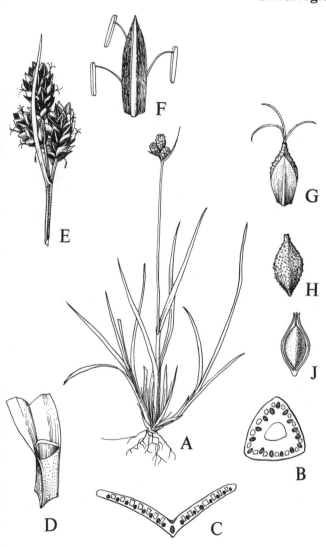

64-69. Carex nigra group

64. C. recta Boott
65. C. aquatilis Wahlenb.
66. C. acuta L.
67. C. nigra (L.) Reichard
68. C. elata All.
69. C. bigelowii Torrey ex Schweinitz

The following six species belong to the section of subgenus *Carex* which has two stigmas; they have characteristic red- or purple-brown/black glumes with green or pale midribs usually ceasing below the apex, and green, biconvex utricles with beak short or almost absent. In the main, their geographical distributions overlap and ecologically they have similar requirements. They form partially fertile hybrids which backcross, leading commonly to introgression between the species, and in all species there are forms showing one or more characters of another species. *C. nigra*, the most ubiquitous (both geographically and ecologically) species of all, has therefore understandably the greatest variation. In the space allocated here justice cannot be done to all these varieties and forms and the figures illustrate the type variety.

C. trinervis Degl. is also in this group; it is a species of dune slacks of W Europe with distinctive inrolled lvs (very similar to those of marram grass) and with at least the lower ♀ glumes with a long excurrent midrib. It has been recorded from E Norfolk from a most unlikely inland habitat and has not been refound; material purporting to be of Norfolk origin, grown first by C. E. Salmon and then A. W. Stelfox and seen in the latter's garden is certainly *C. trinervis* (see p. 250).

The following hybrids within the group have been recorded in the British Isles:

> *C. acuta* × *elata* (=*C.* × *prolixa* Fries; *C.* × *curtisii* Druce nom. nud.)
> *C. acuta* × *nigra* (=*C.* × *elytroides* Fries)
> *C. aquatilis* × *bigelowii* (=*C.* × *limula* Fries)
> *C. aquatilis* × *nigra* (=*C.* × *hibernica* Ar. Benn.)

Bennett (1897) wrongly attributed *C. elata* as a putative parent of *C.* × *hibernica*, and the name is best interpreted as given here, on morphological grounds and on the fact that *C. elata* is not found in the Galway Bridge, Kerry, area whereas *C. nigra* is.

C. aquatilis × recta (=C. × grantii Ar. Benn.)
C. bigelowii × nigra (=C. × decolorans Wimmer)
C. elata × nigra (=C. × turfosa Fries)

C. nigra × recta (=C. × spiculosa Fries), a plant collected in Harris (v.c. 110) and attributed to this hybrid, is in our opinion a form of C. nigra as stated by Wilmott (1938).

C. acuta × aquatilis has been recorded from Europe.

Hybrids with species outside the group include the following:

C. acuta × acutiformis (=C. × subgracilis Druce)
The statement (Jermy, 1967) that this hybrid may be partly fertile needs further investigation as no germination of the fully formed seeds has yet been achieved.

C. bigelowii × binervis was recorded by E. S. Marshall from Ben Wyvis but not substantiated.

C. flacca × nigra (=C. × winkelmannii Ascherson & Graebner) described from Germany has been recorded from time to time from Britain but the plants have usually turned out to be variants of either parent.

C. acuta × vesicaria, C. acutiformis × elata and C. elata × riparia have been recorded from Europe.

Faulkner (1972) discussed the cytology of British and some European members of this group and found similarities in chromosome number and morphology between C. acuta, C. juncella and C. nigra and between C. aquatilis and C. elata. As a result of artificial crossing, the chromosomes of C. nigra were shown to pair completely with those of C. juncella, thus supporting the view of Hylander (1966) that the latter was closely related to C. nigra and is best thought of as a northern variant of that species. It has yet to be found in its extreme form in the British Isles. Further, chromosome studies on a number of populations of C. recta, both in Scotland and Scandinavia, suggest the taxon as construed in Europe may best be regarded as of hybrid origin, although Scottish plants are not identical to those from Scandinavia either in morphology or chromosome number or behaviour. It is suggested that both on plant and chromosome morphology the parents of the Scottish C. recta may be C. aquatilis and C. paleacea Wahlenb. We find a substantial amount of seed set (although germination has yet to be proved) and we believe that introgression or back-crossing with C. aquatilis is prevalent in some localities. For this reason we prefer to treat C. recta as a species with ability to reproduce from seed rather than as a hybrid.

64. Carex recta Boott

Estuarine Sedge Map 59

Rhizomes often far-creeping; shoots tufted; roots pale purple-brown; scales red-brown, often blackened, persistent. *Stems* 30–100 cm, stiff, trigonous. *Lvs* up to 110 cm × 3–6 mm, smooth beneath, rough above and on margins, ± flat or weakly keeled, ± abruptly tapering to a flat, stiff point, mid- to yellow-green; sheaths herbaceous becoming brown or tinged wine-red, persistent, transverse septa distinct, inner face hyaline, ± persistent; ligule 2–4 mm, obtuse, tubular. *Infl.* $\frac{1}{6}$–$\frac{1}{5}$ length of stem; bracts lf-like with purplish auricles, at least the lower exceeding infl. ♂ *spikes* 1–4, 10–40 mm, lower pedunculate; ♂ *glumes* 4–5 mm, oblanceolate-elliptic, margins hyaline; apex obtuse. ♀ *spikes* 2–4, contiguous or ± overlapping, 30–80 mm, lax fld at base, cylindric, upper often ♂ at top; peduncles 1–3 cm, rough; ♀ *glumes* 4–5 mm, ovate-lanceolate; apex acute or acuminate, aristate in lower florets. *Utricle* 2.5–3 mm, obovate or suborbicular, green, faintly nerved; beak c. 0.2 mm, truncate; stigmas 2; nut orbicular or obovate, biconvex. Fr. 8–9.

A very local species of estuarine or lower riverine situations in NE Scotland (E Inverness, E Ross, E Sutherland, Caithness) but there often forming extensive colonies. In stiff peaty alluvium or in more sandy situations but usually where silt is periodically deposited or the water-table fluctuates seasonally. Often a dominant with *Phragmites*, *Phalaris arundinacea* and other grasses in species-rich mire or sometimes forming a pure sedge community with *C. aquatilis*, *C. nigra*, etc.

Distinct in having aristate glumes at the base of each ♀ spike; similar in other ways to *C. aquatilis* and *C. acuta* but it lacks the glaucous upper surface of the leaves of these species and the stem is intermediate between the two.

For hybrids see page 197.

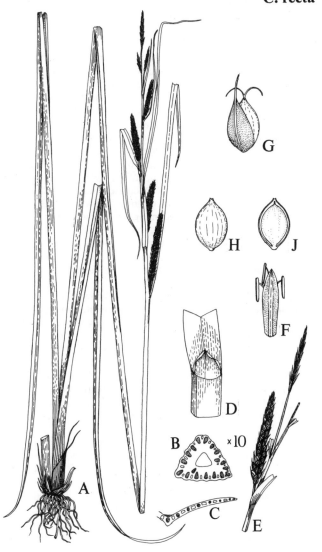

65. Carex aquatilis Wahlenb.

Water Sedge   Map 56

Rhizomes far-creeping; shoots tufted; roots up to 2 mm thick, much branched, red-brown; scales orange- or red-brown, persistent. *Stems* 20–110 cm, smooth, bluntly trigonous to subterete, brittle. *Lvs* 15–100 cm × 3–5 mm, smooth, shiny, dark- or yellow-green beneath, rough, ± glaucous above, flat or ± channelled, gradually tapered to a flat point, the margins rolling inwards on drying; sheaths herbaceous, wine-red or red-brown, persistent, inner face hyaline, soon decaying, apex straight or concave; ligule 5–10 mm, obtuse. ± tubular. *Infl.* $\frac{1}{5}-\frac{1}{4}$ length of stem; bracts lf-like with purplish auricles, lowest exceeding infl. ♂ *spikes* 2–4, 5–50 mm; ♂ *glumes* 3–4 mm, obovate to oblanceolate, margins hyaline; apex obtuse. ♀ *spikes* 2–5, contiguous or lowest ± distant, 20–60 mm, cylindric, slender, dense-fld, becoming lax below, upper ± sessile, often ♂ at top; peduncles 10–20 mm, smooth; ♀ *glumes* 2–3 mm, ovate or broadly elliptic, margin hyaline; apex subacute or obtuse. *Utricles* 2–2.5 mm, ellipsoid to obovoid, nerveless, green; beak almost 0, truncate; stigmas 2; nut ellipsoid to obovoid, biconvex.   Fr. 7–9.

C. aquatilis is a swamp species. In lakes it grows with *C. vesicaria*, *C. rostrata*, etc; in rivers in the N it is often the only sedge in the scanty riparian flora but further south it grows with *C. riparia* and *C. acutiformis*. In E-central Highlands *C. aquatilis* grows with *C. curta*, *C. rariflora*, *C. nigra* and *Sphagnum* spp., above 720 m (2400 ft) in less flushed areas on gently sloping deep peat; or without *C. rariflora* on upland river terraces, often where snow lies late. Basically a northern species from Shetland to the Scottish border, although extinct in many former localities; local in the Lakes and W Wales and scattered in Ireland, where it needs further investigation.

Easily distinguished by the shiny bright green under surface of the leaf, the red sheaths and the slender ♀ spikes with very long bracts and nerveless utricles. Further the brittle, ± subterete stem (which snaps when folded on collection) is characteristic of the species. *C. acuta*, which is similar in stature, has leavesg laucous beneath, rarely red sheaths, more robust spikes with faintly ribbed utricles and tough stems which do not snap.

For hybrids see pages 196–7.

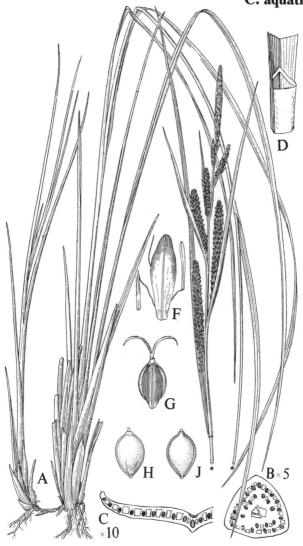

66. Carex bigelowii   Torrey ex Schweinitz

Stiff Sedge   Map 57

Rhizomes shortly creeping; shoots solitary or in pairs, $\pm$ close together; roots grey-brown, often as thick as rhizomes; scales red- or purple-brown, shiny, persistent. *Stems* 4–30 cm, rough towards the top, rigid, thick, sharply trigonous. *Lvs* up to 25 cm $\times$ 2–7 mm, stiff or arcuate, $\pm$ rough, keeled, $\pm$ abruptly tapered to flat apex, margins often revolute and rolling outwards on drying, glaucous, dying to a red-brown; sheaths brown or red-brown and shiny, persistent, inner face hyaline, soon decaying, apex $\pm$ straight; ligule 1–2 mm, acute, shortly tubular. *Infl.* $\frac{1}{8}$–$\frac{1}{5}$ (or in short specimens up to $\frac{1}{2}$) length of stem; bracts lf-like, with blackish auricles, lowest shorter than infl. $\male$ *spike* 1, rarely 2, 5–20 mm; $\male$ *glumes* 3–4 mm, obovate-elliptic; apex acute or rounded. $\female$ *spikes* 2–3, contiguous or lower $\pm$ distant, 5–15 mm, cylindric-ovoid, erect, sessile or lowest only sometimes shortly pedunculate; $\female$ *glumes* 2.5–3.5 mm, ovate-elliptic; apex obtuse. *Utricle* 2.5–3 mm, broader than glumes, suborbicular-ellipsoid, green; beak c. 0.2 mm, truncate or slightly notched; nut ovoid-ellipsoid, biconvex.   Fr. 7–8.

C. bigelowii is a mountain plant occurring above 600 m (2000 ft) altitude for most of its range. A dominant sedge in the *Rhacomitrium-Vaccinium* lichen-rich heath or with *Dicranum fuscescens* or *Polytrichum alpinum* in areas of deep snow cover. In flushed gullies it can become a robust plant. In N Wales, N Pennines and Lake District, S Uplands and W Highlands; in Ireland a plant of mountain summits and ridges.

The shiny red-purple-brown rhizome scales distinguish this from *C. nigra* and from *C. binervis* which has orange-brown scales and a dark green, not glaucous, leaf.

For hybrids see pages 196–7.

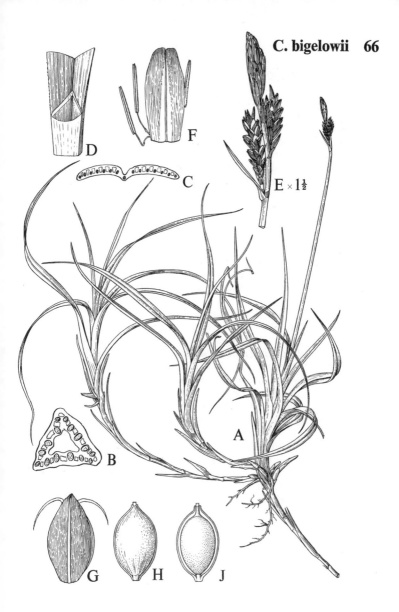

E × 1½

67. Carex elata All.

Tufted Sedge

Map 59

Rhizomes very short, erect; shoots densely tufted, forming tussocks up to 40 cm high; roots up to 2 mm thick, purple-brown; scales light brown, shiny, persistent. *Stems* 25–100 cm, rough, sharply trigonous, solid. *Lvs* 40–100 cm × 3–6 mm, rough, thin, plicate, gradually tapering to a flat apex, glaucous, the margins rolling outwards on drying; sheaths becoming yellow-brown, persistent, inner face hyaline, fibrillose on splitting, apex concave; ligule 5–10 mm, acute, $\pm$ tubular. *Infl.* c. $\frac{1}{7}$ length of stem; bracts lf-like to setaceous, lowest not half length of infl. ♂ *spikes* 1–3, 15–50 mm, lowermost occasionally ♀ at base; ♂ *glumes* c. 5 mm, oblanceolate, margin hyaline; apex obtuse. ♀ *spikes* 2–3, usually contiguous, 15–40 mm, cylindric, erect, $\pm$ sessile, often ♂ at top; ♀ *glumes* 3–4 mm, ovate-elliptic, margins hyaline; apex obtuse, or subacute. *Utricles* 3–4 mm, broadly ovoid-ellipsoid, ribbed, green; beak 0.2 mm, truncate; nut obovoid, biconvex, stalked.   Fr. 5–6.

C. elata is found in eutrophic mire where there is at least seasonal flooding and is therefore common by fen ditches, rivers and lakes. An important component of E Anglian reed-swamp with *C. acutiformis*, *C. riparia*, *Cladium* and *Phragmites*, in places preceding bush colonisation. Also in upland fens with *C. rostrata*, *C. nigra*, *Phragmites* etc. Mainly in E England and central Ireland where base-rich mires are common; rare in Scotland.

A variable species but distinct amongst those in this group in completely lacking leading rhizomes and in having fibrillose sheaths. Depauperate specimens (e.g. in drained ponds) that form tufts rather than tussocks are difficult to separate from non-fruiting *C. acutiformis* but this has red-tinged sheaths and long rhizomes and a more yellow-green leaf upper surface. A very similar species, *C. cespitosa* L. of N and C Europe (confused with *C. caespitosa* Gay [= *C. elata*] and erroneously recorded for Britain) has red-brown scales.

For hybrids see pages 196–7.

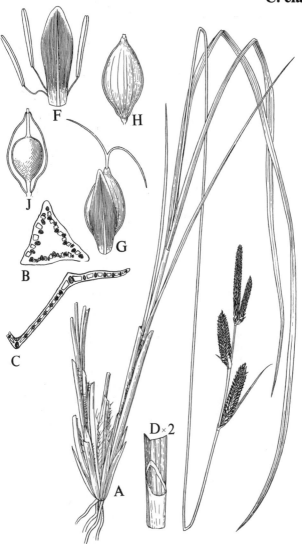

68. Carex nigra (L.) Reichard

Common Sedge Map 60

Rhizomes far-creeping (but see below); shoots tufted; roots pale or red-brown; scales brown or red-brown, $\pm$ shiny, persistent or fibrous. *Stems* 7–70 cm, rough above, slender, trigonous, solid. *Lvs* up to 90 cm $\times$ 1–3 (–5) mm, thin, $\pm$ flat, gradually tapering to a fine point, glaucous, the margins rolling inwards on drying; sheaths brown, black or rarely red, $\pm$ fibrous, inner face hyaline, apex straight; ligule 1–3 mm, rounded. *Infl.* $\frac{1}{6}$–$\frac{1}{4}$ length of stem; bracts lf-like, lowest $\pm$ equalling infl. ♂ *spikes* 1–2, 5–30 mm; ♂ *glumes* 3–5 mm, obovate-oblong; apex obtuse or subacute. ♀ *spikes* 1–4, $\pm$ contiguous, 7–50 mm, cylindric, erect, upper often ♂ at top, lower sometimes distant, with short peduncle; ♀ *glumes* 2.5–3.5 mm, lanceolate-oblong, margin narrowly hyaline; apex obtuse to acute. *Utricles* 2.5–3.5 mm, broader than glumes, ovoid-ellipsoid, faintly ribbed, green, often tinged dark purple; beak almost 0, truncate; nut ellipsoid, biconvex. Fr. 6–8.

C. nigra shows a wide habitat tolerance but is usually in mires or bogs with some degree of water movement or mineral (especially Ca and Mg) enrichment. In upland areas the commonest sedge in flushes with *C. panicea*, *C. demissa*, *C. echinata*, *Molinia*, *Juncus effusus*, *Sphagnum* spp. etc. Similarly in lowland marshes, dune-slacks and stream-sides. Throughout the British Isles.

A very variable species and responsive to habitat conditions. Kükenthal (1909) lists ten varieties (under *C. goodenowii* Gay) and many forms of each which are the result of habitat influence or crossing with other spp. of this group (see pp. 196–7). Dune-slack forms and plants growing around hill sheep-shelters react (to higher phosphate of soil?) in having short, wide, often arcuate leaves (= *C. goodenowii* var. *stolonifera*). Those in mineral-poor mires have slender channelled leaves (= var. *strictiformis*); a tufted form of calcareous mires (var. *tornata*) has wide, rigid leaves and a thick dense spike. Of the varieties that show introgression, var. *recta* has the long bracts and red leaf-sheaths of *C. aquatilis*. Var. *juncea* (*C. juncella* (Fries) Th. Fries) has been recorded from stagnant mires in error for var. *subcaespitosa* (a tussocky growth form); var. *juncea* is a N Scandinavian tussock-form with no long rhizomes and orange-brown basal sheaths.

For hybrids see page 196–7.

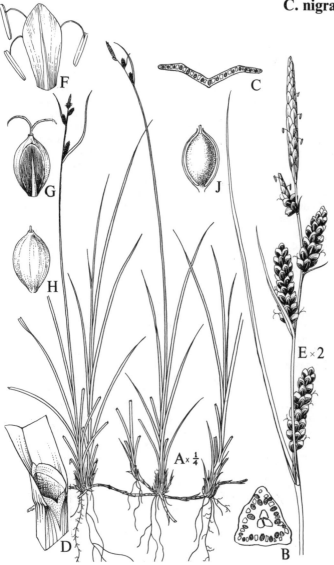

69. Carex acuta   L.

Slender Tufted Sedge   Map 58

Rhizomes far-creeping; shoots tufted; roots brown or red-brown; scales red-brown, often soon decaying. *Stems* 30–120 cm, rough and sharply trigonous above, subterete at base. *Lvs* 30–140 cm × 3–10 mm, rough on edges, thin, plicate, gradually narrowed to a pendulous tip, glaucous, the margins rolling outwards on drying; sheaths brown or red-brown, persistent, transverse septa prominent, inner face hyaline, persisting as a brown membranous strip, apex straight or concave; ligule 4–6 mm, obtuse, truncate. *Infl.* $\frac{1}{6}$–$\frac{1}{4}$ length of stem; bracts lf-like, lowest exceeding infl. ♂ *spikes* usually 2–4, 20–60 mm; ♂ *glumes* 4.5–5.5 mm, elliptic- to obovate-oblong; apex obtuse or subacute. ♀ *spikes* 2–4, usually contiguous, 30–100 mm, cylindric, often lax-fld at base, erect, upper sessile, often ♂ at top, lower shortly pedunculate; ♀ *glumes* 2.5–4 mm, oblong-obovate; apex obtuse or margin inrolled forming a cusp. *Utricles* 2–3.5 mm, ellipsoid-obovoid to subglobose, either longer or shorter than glume, faintly ribbed, green; beak almost 0, truncate; stigmas 2; nut obovoid, biconvex, shortly stalked.   Fr. 6–7.

C. acuta is a species of ponds, dykes and riversides or of marshy places where there is a ± constantly high water level, associated with *C. acutiformis, C. riparia, C. elata,* etc. Mainly in the S and W, becoming rare in Scotland; in Ireland, local in the N and E.

Similar to *C. acutiformis* in leaf-texture and infl. but the three stigmas and serrulate acumen of the ♀ glume of that species is usually sufficient to separate them. If not flowering, the fibrillose leaf-sheaths of *C. acutiformis* and the dark more yellow-green upper surface to the leaf distinguish that species. *C. acuta* can be confused with *C. aquatilis* where their distribution overlaps (see p. 200). A robust form of *C. acuta* of open water habitats has, at the base of the lowermost ♀ spike, glumes with long excurrent midribs; this should not be confused with *C. recta* which has a blunter angled stem and non-glaucous lvs.

C. acuta and *C. acutiformis* hybridise (= *C.* × *subgracilis* Druce) resulting in intermediate sterile offspring with either two or three stigmas in the same spike; for other hybrids see pages 196–7.

208

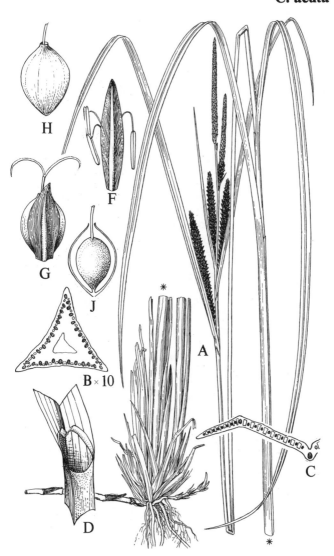

70. Carex microglochin Wahlenb.

Bristle Sedge   Map 6

Rhizomes slender, shortly creeping; shoots often single; roots pale yellow-brown; scales pale brown, persistent. *Stems* 5–12 cm, stiff, erect, solid, subterete, striate. *Lvs* 1–5 cm × 0.5–1 mm, thick, stiff, erect, channelled, ± truncate and rounded at apex, mid-green; sheaths soon becoming brown and decaying, rarely fibrous, inner face hyaline becoming brown, apex straight, serrate; ligule c. 0.5 mm, rounded, tubular. *Infl.* a single few-fld, terminal *spike* 3–5 mm, ♂ above, ♀ below; bracts 0. ♂ *glumes* 2.5–3 mm, ovate-lanceolate, red-brown, with pale midrib; apex ± acute. ♀ *glumes* c. 2 mm, ovate-lanceolate, red-brown, with pale margin, caducous, apex hyaline, acute. *Utricles* 3.5–4.5 (–6) mm, narrowly conical, rounded at base, faintly ribbed, pale yellow-green becoming straw colour, apex tapered into beak c. 1 mm long, deflexed at maturity; a stiff bristle, arising from the base of the nut, protrudes from the top of the beak; stigmas 3; nut cylindric-trigonous.   Fr. 7–9.

C. microglochin is a member of a distinctive high level facies of *Carex-Saxifraga aizoides* flush in which other alpine rarities occur (e.g. *C. atrofusca*, *Juncus biglumis*, *J. castaneus* etc) together with *C. demissa*, *C. dioica*, *Thalictrum alpinum* and the moss *Blindia acuta*. In gently sloping, stony, micaceous flushes between 600–900 m (2000–3000 ft) where the total plant cover is usually less than 50 per cent. First found in 1923 in the Breadalbanes, Perthshire, its only British station; the record for Harris is in error.

This species could be confused with *C. pauciflora*, although the habitats are distinct. It is distinctive in having an exserted bristle arising from well below the nut but within the utricle wall whose outward appearance is very similar to and should not be mistaken for the persistent style of *C. pauciflora*. Also the utricles are fewer on each infl. and, on the average, smaller than in *C. pauciflora*. The shoots of the latter species are loosely clustered, not solitary or 2 or 3, and the lower leaf-sheaths have a short subulate lamina, 1–2 mm long, a feature not seen in *C. microglochin*.

No hybrids of this species are known.

Figs. D[1]: side view of ligule; K: section of utricle showing origin of bristle

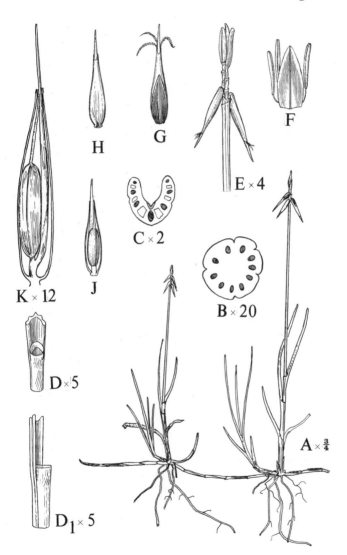

71. Carex pauciflora Lightf.

Few-flowered Sedge Map 5

Rhizomes slender, shortly creeping, often much branched, forming an open mat; shoots loosely tufted; roots cream or pale yellow-brown; scales pale brown (rarely dark), striate, persistent. *Stems* 7–27 cm, stiff, solid, trigonous, often curved. *Lvs* up to 20 cm × 1–2 mm, with c. 9 veins, stiff, thick, ± channelled, mid-green, gradually narrowed to a wide, rounded apex, those of sterile shoots often narrower and setaceous; sheaths pink- (rarely red-) brown, persistent, some with very short subulate green blades only, inner face hyaline, apex straight; ligule c. 0.5 mm, rounded, tubular. *Infl.* a single, few fld, terminal *spike* 3–8 mm, ♂ above, ♀ below; bracts 0. ♂ *glumes* 3.5–5 mm, lanceolate, pale red-brown with hyaline margins; apex ± acute. ♀ *glumes* 3.5–4.5 mm, broadly lanceolate, clasping utricle, caducous, pale red-brown, margins hyaline towards the ± acute apex. *Utricles* 5–7 mm, subfusiform, tapered more abruptly below, faintly nerved, pale yellow-, rarely red-brown, tapered above to a beak-like apex; stigmas 3; style persistent in fruit, protruding from apex of utricle; nut oblong-cylindric, trigonous. Fr. 6–7.

C. pauciflora is a species of oligotrophic bogs characterised by *Sphagnum* spp., *Eriophorum* spp., *Erica tetralix*, *Narthecium ossifragum* etc. Throughout the Highlands and W Scotland becoming less frequent to the E and scattered in the Outer Isles; not recently recorded in Orkney and absent from Shetland. In England in the Lake District and N Pennines with one isolated locality in NE Yorks and one in Snowdonia; in Ireland only in Antrim.

Confused with *C. pulicaris*, although in the field their ecological preferences keep them distinct. *C. pulicaris* is always easily distinguished by the fine red-brown roots and the darker brown ellipsoid utricle which lacks the persistent style of *C. pauciflora*. This style should not be confused with the bristle of *C. microglochin* (see p. 210). *C. dioica* is vegetatively similar but has a narrower, channelled and inrolled leaf with only three veins.

No hybrids of *C. pauciflora* are known.

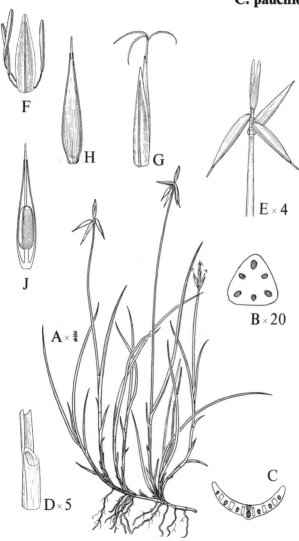

72. Carex rupestris   All.

Rock Sedge   Map 2

Rhizomes shortly creeping, often much branched; shoots ± tufted; roots purple-brown, slender; scales shiny, red-brown, persistent. *Stems* 7–20 cm, stiff, solid, trigonous. *Lvs* 5–20 cm × 1–1.5 mm, often curled or twisted, flat below, becoming keeled towards the gradually attenuated, trigonous apex, dull dark green, persistent on dying; sheaths red- or orange-brown, ribbed, persistent, inner face hyaline, becoming brown and persisting, apex brown, straight; ligule 1–2 mm, rounded, ± tubular. *Infl.* a single, terminal *spike* 7–15 mm, ♂ above, ♀ below; bracts glumaceous with setaceous points, caducous. ♂ *glumes* 2.5–3 mm, ovate, red- or purple-brown, with obscure midrib; apex acute. ♀ *glumes* 2.5–3.5 mm, elliptic to broadly ovate, dark red- or purple-brown, with obscure midrib and narrowly hyaline margin, persistent; apex obtuse, sometimes mucronate. *Utricles* c. 3 mm, obovoid, trigonous, faintly ribbed below, grey-green to brown, erecto-patent to erect at maturity; beak c. 0.3 mm, slightly notched; stigmas 3; nut broadly ellipsoid, trigonous.   Fr. 7–8.

A local plant of limestone cliff-ledges or on siliceous rock-ledges where influenced by calcareous flushing; often on unstable slopes. Associated with general flush species e.g. *C. flacca*, *C. demissa*, *C. pulicaris*, *Festuca ovina* etc; occurs in N Scotland in *Dryas octopetala* dwarf shrub heath. In Scotland only; usually above 600 m (2000 ft) in the Grampians, much lower in W Ross and W Sutherland, where with *C. capillaris* it descends to sea level. A record from a hill in S Uist, where no base-rich rock exists, is almost certainly an error.

C. rupestris can be confused with *C. pulicaris*, with which it often grows. The latter species has narrow straighter leaves, V-shaped (keeled) in transverse section, and the leaf-sheaths are fibrous and brown, not red-brown as in *C. rupestris*. The young infl. looks similar, but *C. pulicaris* has only two stigmas and the ♀ glume has a broader hyaline margin.

No hybrids of *C. rupestris* are known.

214

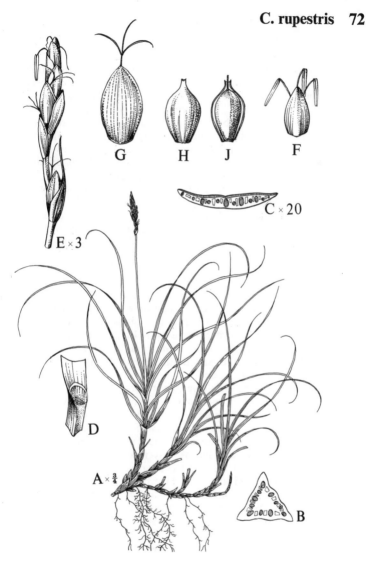

E × 3

G

H

J

F

C × 20

D

A × ¾

B

73. Carex pulicaris L.

Flea Sedge   Map 4

Rhizomes shortly creeping; shoots often densely tufted; roots very fine, red-brown; scales pale or purplish-brown often becoming fibrous. *Stems* 10–30 cm, slender, stiff, terete. *Lvs* 5–25 cm × 0.5–1 mm, with c. 9 veins, ± stiff, keeled, dark green, apex blunt; sheaths becoming brown, ribbed, lower fibrous or soon decaying, inner face hyaline, apex ± straight; ligule usually less than 0.5 mm, rounded. *Infl.* a single, few-fld, terminal *spike* 10–25 mm, ♂ above, ♀ below; bracts glumaceous. ♂ *glumes* 4.5–5 mm, oblong-elliptic, purple- or rarely red-brown, with paler margin; apex obtuse or ± acute. ♀ *glumes* 3.5–4 mm, caducous, broadly lanceolate, red- or purple-brown, midrib keeled, margins sometimes narrowly hyaline; apex acute or ± obtuse. *Utricles* 4–6 mm, ellipsoid or oblanceolate, shortly and stoutly pedicelled, dark brown, shiny, deflexed on maturity; beak 0.2 mm, slightly notched; stigmas 2; nut narrowly obovoid, biconvex, tapered to thick stalk.   Fr. 6–7.

C. *pulicaris* is a species of meso- to eutrophic silty soils with impeded drainage or on wet slopes where drainage water comes from calcareous rocks. A common constituent of the *Carex*-hypnoid moss mires on the mica-schists and similar calcareous rocks of C and NW Scotland; in S Uplands, N England and N Wales more often found in species-rich *Nardus* grassland. In valley-bog and calcareous mires of E and S England; scattered in a variety of habitats throughout Ireland.

C. *pulicaris*, when in flower, i.e. before the utricles are deflexed, may be mistaken for C. *rupestris* with which it often grows; see p. 214. When not in flower, it may be confused with two other species with narrow leaves, C. *dioica* and C. *pauciflora*. The leaf of the former is three-veined and inrolled, while the latter grows mainly in oligotrophic mires and has seven-veined leaves (like C. *pulicaris*) which are c. 2 mm wide and channelled. The leaf-sheaths of C. *pulicaris* are ribbed and deep brown; those of C. *dioica* and C. *pauciflora* are pale and barely striate. The fine red-brown roots of C. *pulicaris* are always distinctive. Further the mature utricles sometimes do not reflex and then simulate a ♀ C. *dioica*, but the utricle shape and lack of ribs distinguish it.

No hybrids of C. *pulicaris* are known.

216

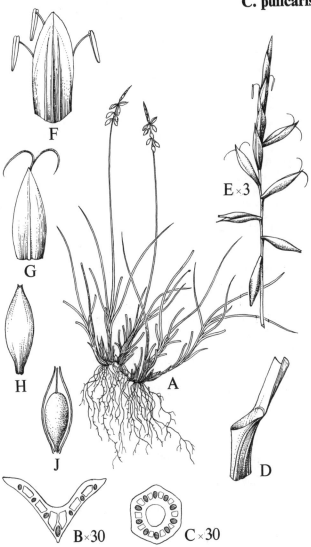

F

E×3

G

H

J

A

D

B×30   C×30

NOTE ON THE MAPS

The majority of the maps have been provided by the Biological Records Centre of the Institute of Terrestrial Ecology to whom grateful acknowledgement is made. These have been updated since the publication of the *Atlas* and the division between older and newer records has been advanced to 1950.

The maps for species 2, 4, 11, 13, 18, 19, 28, 29, 32, 36, 37, 41, 44, 49–51, 53–55, 57, 59–64, 70 and 72 have been newly researched by R.W.D.; only those records have been accepted which have been confirmed in the field, or by voucher specimens in herbaria, or, in a very few cases, by other testimony that has been considered incontrovertible.

The maps for two difficult groups—the *C. muricata* aggregate (*C. spicata*, the two subspecies of *C. muricata*, and the two subspecies of *C. divulsa*), and the two taxa sometimes recognised in *C. serotina*—are selective. They include only those plants positively determined, either in the field or in herbaria, in the first instance by R.W.D., in the second by A.O.C. and R.W.D. working together. The maps are therefore incomplete, for the plants of each complex no doubt occur in many other localities that have not been checked, and intermediates have been neglected. Nevertheless the selection gives, we believe, a fair representation of the areas in the British Isles where these taxa occur, and are widespread or scarce. Although, in our revision of the text, we have come to the conclusion that to maintain *C. scandinavica* E. W. Davies, even as a subspecies of *C. serotina* as in Flora Europaea, is of little practical value to British botanists, we feel the maps of the extremes of the morphological range are of interest and worthy of publication.

We wish to emphasise that the maps are not intended to pin-point localities, but to give an overall impression of the distribution more striking than can be conveyed by a verbal description.

The symbols ● and ⊙ stand for records made later than 1949, the symbols o and ⊚ for earlier records not confirmed since that date. The distinction demonstrates how certain species (e.g. *C. echinata*, *C. elongata*, *C. maritima* and *C. pseudocyperus*) have declined through changes of land-use. This change is even more rapid over the period 1950–1970 and had we used the latter date as our time threshold the differences would have been more marked. Local Flora work in Kent, for example, in 1971–79 (E. G. Philp, pers. comm.) shows that for 27 species, plants have not been found in 169 10-km squares in spite of often thorough searching, yet are shown on our maps as present in 1950–1970.

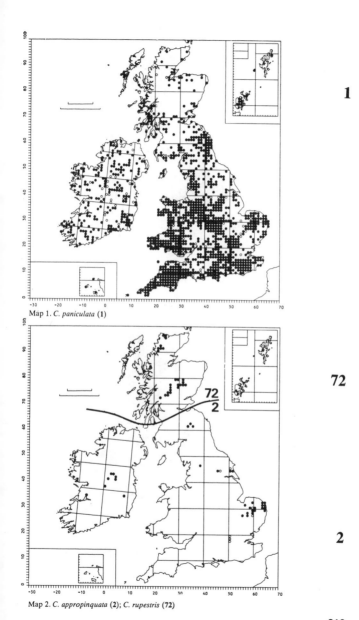

Map 1. *C. paniculata* (**1**)

Map 2. *C. appropinquata* (**2**); *C. rupestris* (**72**)

1

72

2

3

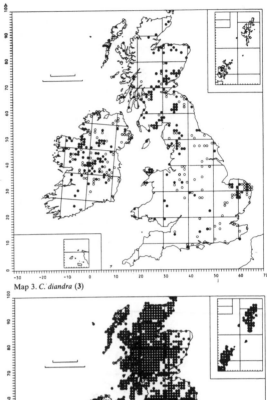

Map 3. *C. diandra* (3)

73

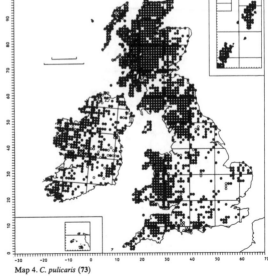

Map 4. *C. pulicaris* (73)

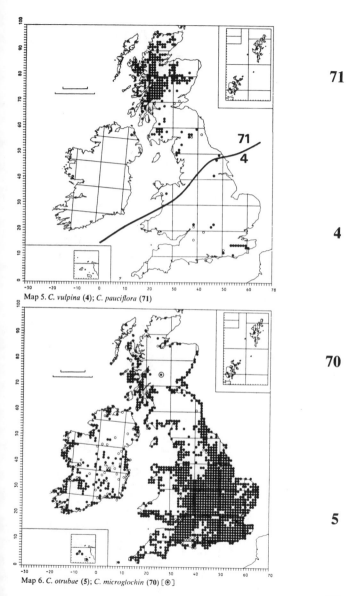

71

4

Map 5. *C. vulpina* (**4**); *C. pauciflora* (**71**)

70

5

Map 6. *C. otrubae* (**5**); *C. microglochin* (**70**) [⊙]

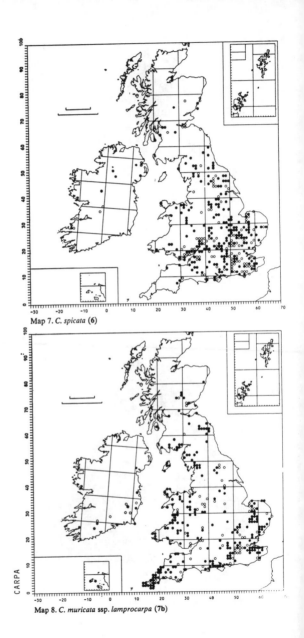

6

7b

Map 7. *C. spicata* (6)

Map 8. *C. muricata* ssp. *lamprocarpa* (7b)

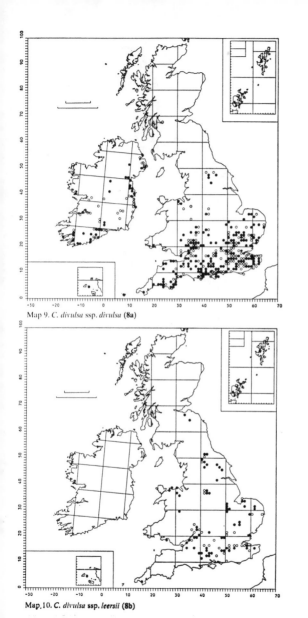

Map 9. *C. divulsa* ssp. *divulsa* (8a)

Map. 10. *C. divulsa* ssp. *leersii* (8b)

8a

8b

9

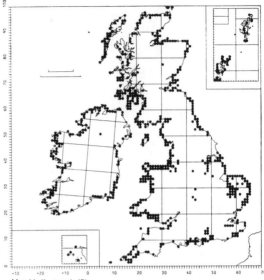

Map 11. *C. arenaria* (**9**)

13

7a

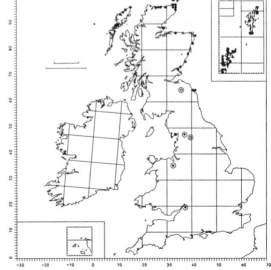

Map 12. *C. muricata* ssp. *muricata* (**7a**) [⊙ ⊚]; *C. maritima* (**13**)

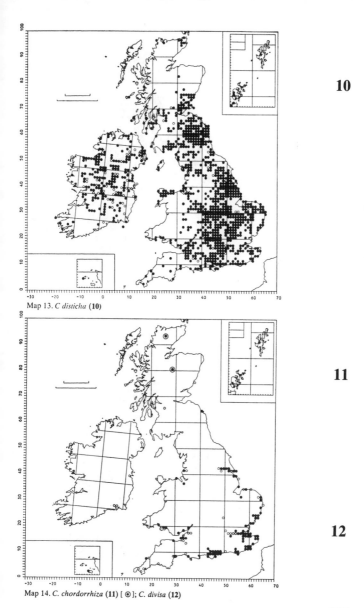

Map 13. *C disticha* (**10**)

Map 14. *C. chordorrhiza* (**11**) [⊙]; *C. divisa* (**12**)

10

11

12

14

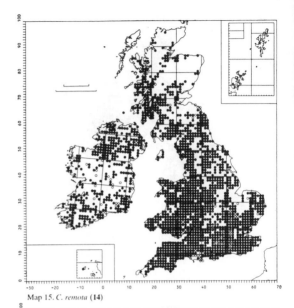

Map 15. *C. remota* (**14**)

15

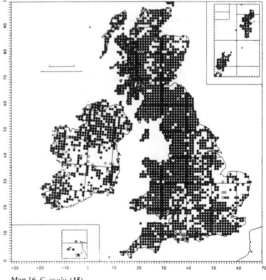

Map 16. *C. ovalis* (**15**)

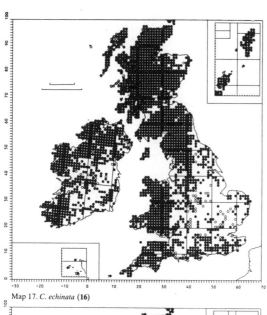

Map 17. *C. echinata* (**16**)

16

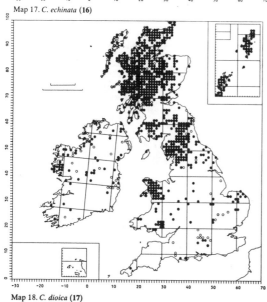

Map 18. *C. dioica* (**17**)

17

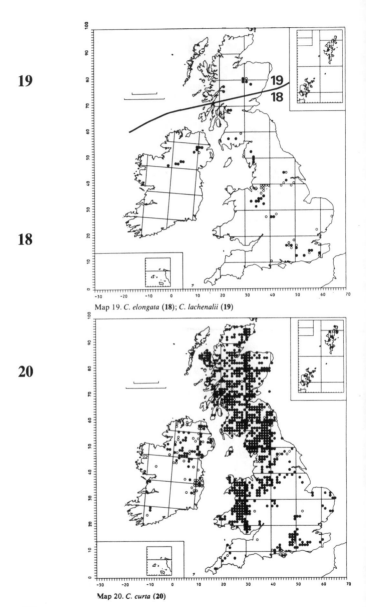

19

18

Map 19. *C. elongata* (**18**); *C. lachenalii* (**19**)

20

Map 20. *C. curta* (**20**)

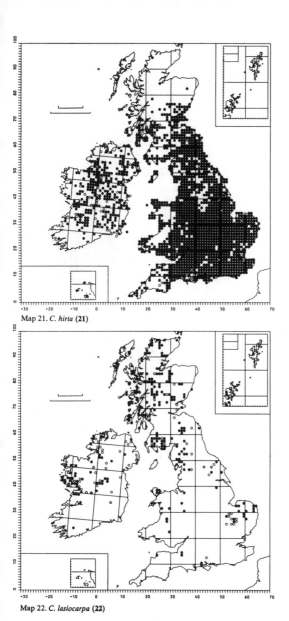

21

Map 21. *C. hirta* (21)

22

Map 22. *C. lasiocarpa* (22)

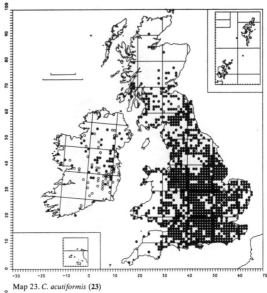

Map 23. *C. acutiformis* (**23**)

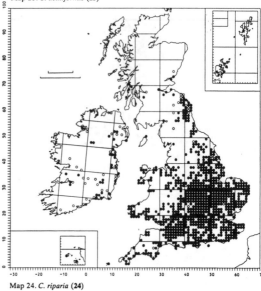

Map 24. *C. riparia* (**24**)

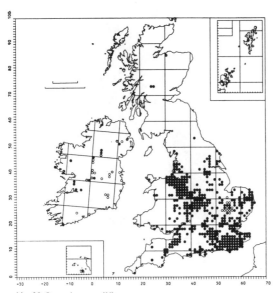

Map 25. *C. pseudocyperus* (**25**)

25

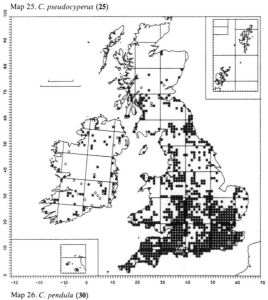

Map 26. *C. pendula* (**30**)

30

26

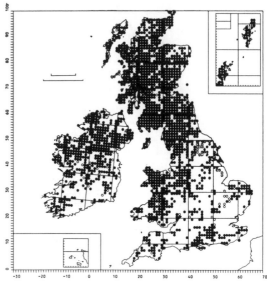

Map 27. *C. rostrata* (**26**)

27

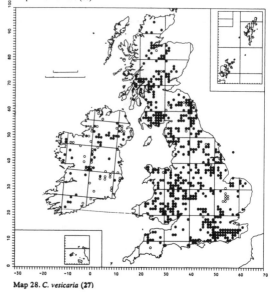

Map 28. *C. vesicaria* (**27**)

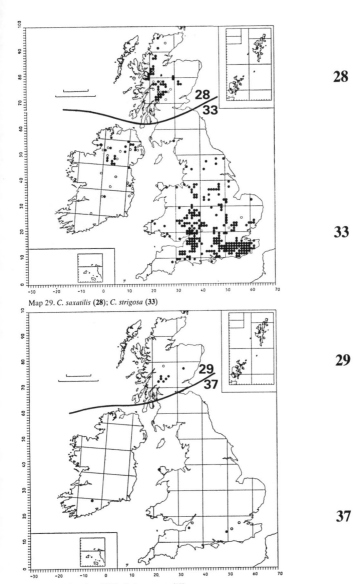

28

33

Map 29. *C. saxatilis* (**28**); *C. strigosa* (**33**)

29

37

Map 30. *C.* X *grahamii* (**29**); *C. depauperata* (**37**)

31

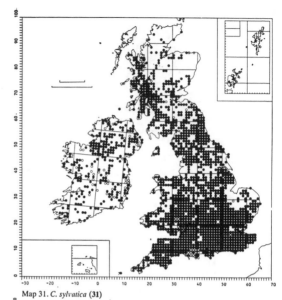

Map 31. *C. sylvatica* (**31**)

32

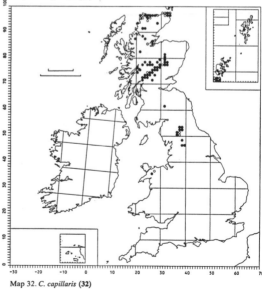

Map 32. *C. capillaris* (**32**)

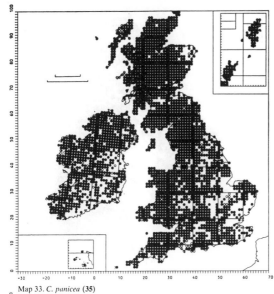

35

Map 33. *C. panicea* (**35**)

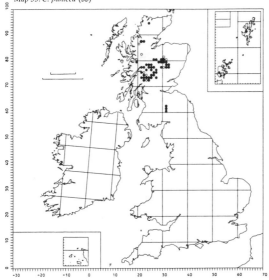

36

Map 34. *C. vaginata* (**36**)

34

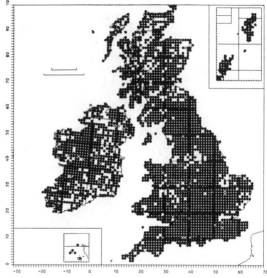

Map 35. *C. flacca* (**34**)

38

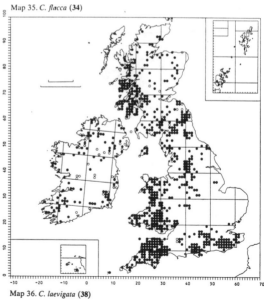

Map 36. *C. laevigata* (**38**)

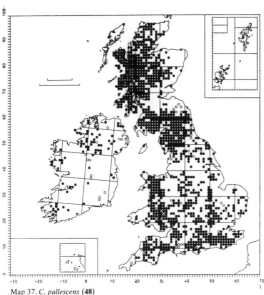

48

Map 37. *C. pallescens* (**48**)

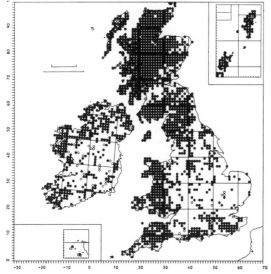

39

Map 38. *C. binervis* (**39**)

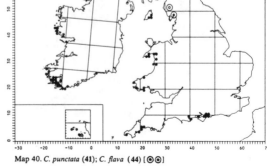

40

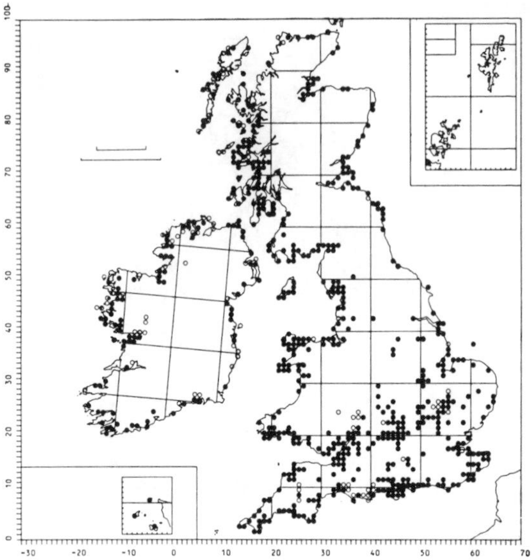

Map 39. *C. distans* (**40**)

44

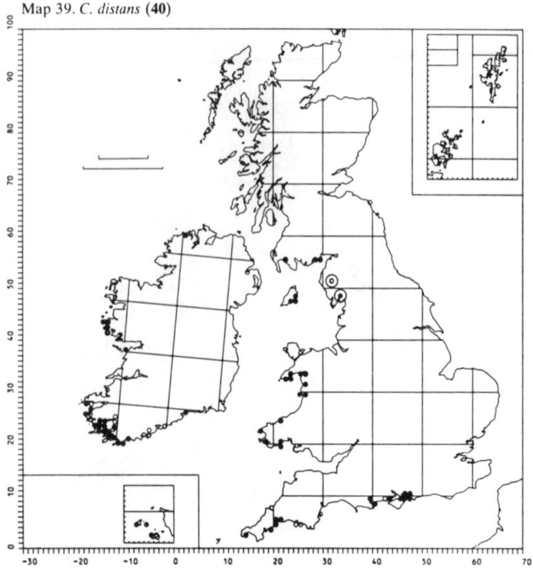

41

Map 40. *C. punctata* (**41**); *C. flava* (**44**) [⊙⊙]

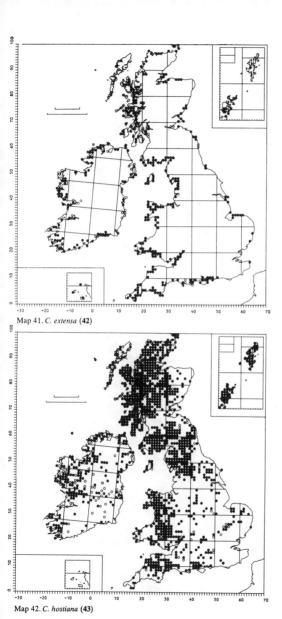

42

Map 41. *C. extensa* (**42**)

43

Map 42. *C. hostiana* (**43**)

45

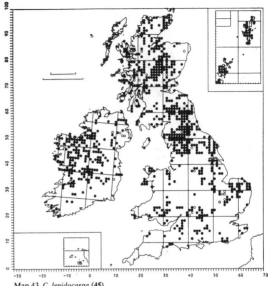

Map 43. *C. lepidocarpa* (**45**)

46

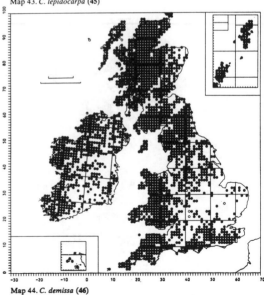

Map 44. *C. demissa* (**46**)

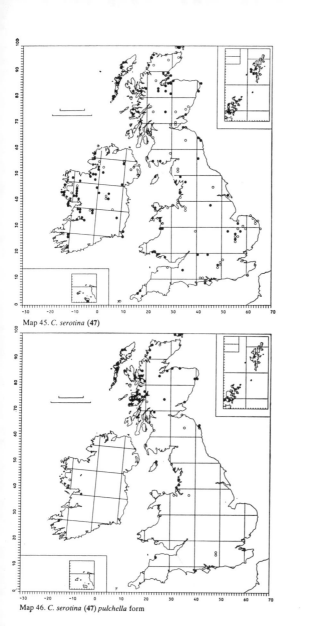

Map 45. *C. serotina* (**47**)

Map 46. *C. serotina* (**47**) *pulchella* form

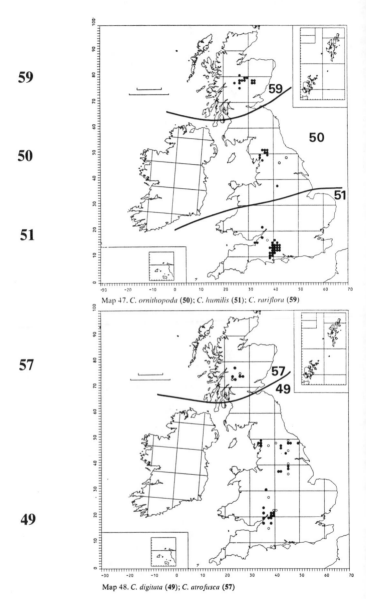

59

50

51

Map 47. *C. ornithopoda* (**50**); *C. humilis* (**51**); *C. rariflora* (**59**)

57

49

Map 48. *C. digitata* (**49**); *C. atrofusca* (**57**)

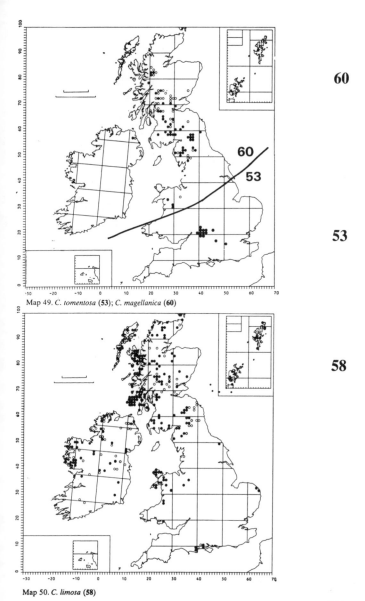

60

53

58

Map 49. *C. tomentosa* (**53**); *C. magellanica* (**60**)

Map 50. *C. limosa* (**58**)

243

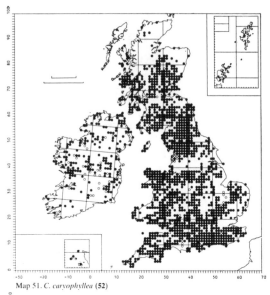

52

Map 51. *C. caryophyllea* (**52**)

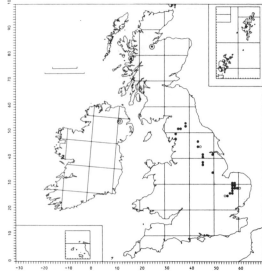

62

54

Map 52. *C. ericetorum* (**54**); *C. buxbaumii* (**62**) [⊙◎]

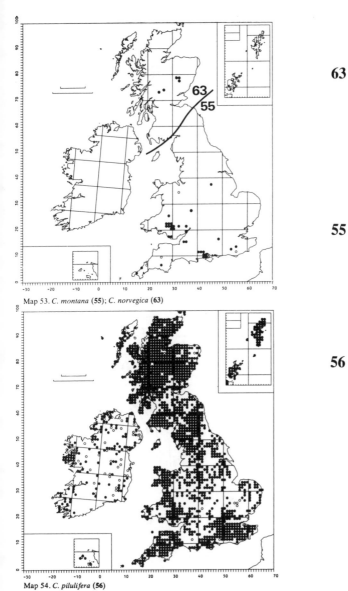

63

55

56

Map 53. *C. montana* (**55**); *C. norvegica* (**63**)

Map 54. *C. pilulifera* (**56**)

61

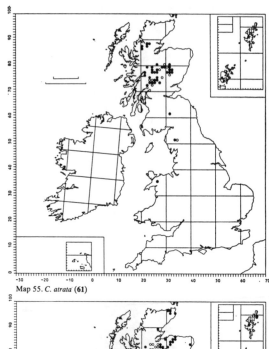

Map 55. *C. atrata* (**61**)

65

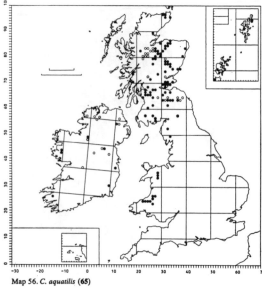

Map 56. *C. aquatilis* (**65**)

246

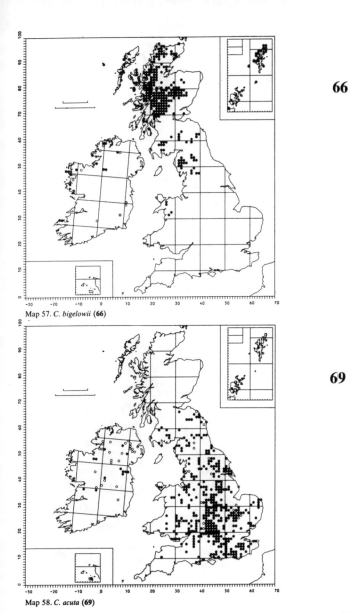

66

Map 57. *C. bigelowii* (**66**)

69

Map 58. *C. acuta* (**69**)

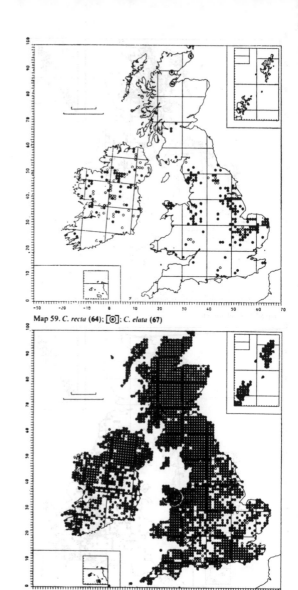

64

67

68

Map 59. *C. recta* (**64**); ⊡ : *C. elata* (**67**)

Map 60. *C. nigra* (**68**)

ALIEN AND EXTINCT SPECIES AND DUBIOUS RECORDS

Alien species

Few alien species of *Carex* have been recorded in the British Isles. Two Australian sedges, *C. inversa* R.Br. (subgenus *Vignea*) and *C. longifolia* R.Br. (subgenus *Carex*) have been found with wool shoddy, the first species in Yorkshire and the Home Counties and the latter in Hampshire. Two eastern N American species (of the subgenus *Vignea*), *C. crawfordii* Fernald and *C. vulpinoidea* Michx, possibly introduced with fodder or other seed, were established in Kent and Surrey for many years, but have now disappeared. The latter is also recorded for Hampshire (1968) and for Dumbarton (1980) (A. G. Stirling, pers. comm.)

C. longifolia R. Br. A tufted plant with *stems* often up to 75 cm; *lvs* 40 cm × 2–5 mm, sheaths brown, shiny, fibrous. *Infl.* of 4–7 distant or remote spikes (often 2 per node) on long (10–15 cm) peduncles; *spikes* 15–25 mm, slender; ♀ *glumes* 4 mm, brown-hyaline, aristate or mucronate; *utricles* 6 mm, obovoid-ellipsoid, ribbed, rough, tapered to fine beak 1.5 mm long.

C. crawfordii Fernald. A tufted species resembling *C. ovalis* (see p. 96) but differing as follows: *Lvs* usually about equalling stems; *spikes* 7–15; ♀ *glumes* 2.5–3 mm. somewhat darker, brown-hyaline; *utricles* 4–5 mm, ellipsoid-lanceolate.

C. inversa R. Br. A shortly creeping plant; *stems* 30–40 cm, slender, acutely trigonous; *lvs* 20–35 cm × 0.75–1.5 mm. *Infl.* a terminal cluster of 2–3 short spikes, ♂ at base, ♀ above, with long overtopping bracts; ♀ *glumes* 2 mm, hyaline or pale brown with green midrib, attenuate; *utricles* 3 mm, ovate, green, ribbed, with rough beak 1 mm long.

C. vulpinoidea Michx. A tufted plant with *stems* 30–90 cm; *lvs* often longer than stem, 2–4 mm wide. *Infl.* spikelike, narrow, much branched but branches short; *glumes* pale brown, aristate, giving 'bristly' appearance to infl.; *utricles* 2 mm, ovate, plano-convex; beak short.

Dubious records

Dandy gives three species with ‡ (dubious records), namely *C. glacialis* Mackenzie, *C. bicolor* All. and *C. capitata* L. The first two were recorded for Rhum, v-c. 104 (Heslop-Harrison, 1941 & 1946) and the last mentioned species for S Uist in the Outer Hebrides, v-c. 110 (Heslop-Harrison, 1946) as a single tuft. All have since disappeared from these localities and we consider them to have been planted. *C. bicolor*, a N European species of a very special habitat, is unlikely to occur in Britain. The occurrence of *C. capitata* and *C. glacialis* is feasible (see below).

Sir J. E. Smith (1830) mentions that he had been sent a specimen by T. Drummond "from near Panmure, Forfar", which Smith identified as *C. secalina* sensu Schkuhr and which was later recorded by Babington (1843) and others as *C. hordeiformis* Wahlenb. (= *C. hordeistichos* Vill.). This is a S and W European species likely to have been introduced by G. Don, whose garden Drummond eventually succeeded to. Another W European species, *C. brizoides* L., recorded in a later edition of Babington (1851) from Studley Wood, Yorks., and not found since, must also remain dubious.

Extinct species

C. davalliana Sm. grew on a calcareous mire at Lansdown, near Bath, until the 1830s but has since been lost through drainage. It occurs through much of Central Asia and C and S Europe from the Altai to N France. It is a densely tufted plant 15–25 cm high, similar to *C. dioica. Lvs* setaceous, rough. ♀ *spikes* 15–20 fld; ♀ *glumes* dark brown with hyaline edges; *utricle* 4–5 mm, dark brown, faintly ribbed (see p. 101, *Fig.* H³).

C. trinervis Degl. (see p. 196) was found at Ormesby, E Norfolk, by H. G. Glasspoole in 1869. It is distinguished from *C. nigra* and *C. acuta* chiefly in having channelled greyish- or glaucous-green leaves, the lower bract channelled and exceeding the inflorescence, 3-veined ♀ glumes, the lower one on the spike aristate, and utricles 3.5–5 mm. It is a plant of damp coastal sand slacks and heaths and the Ormesby area is most atypical. Herbarium material is variable and the species was probably hybridizing with *C. nigra*, as much material labelled *C. trinervis* is closer to that species. Living material found its way into the garden of C. E. Salmon who sent some to Arthur W. Stelfox, then in Dublin, who had the plant alive in his garden in County Down for many years; that plant had the aristate female glumes and inrolled leaf of *C. trinervis*. A thorough search of Ormesby Common has failed to reveal it in recent years, but it may well occur elsewhere.

250

Potential British species

Among the species which might be expected to occur in the British Isles but have not so far been detected, four deserve mention:

C. glareosa Wahlenb. is widespread on the coasts of N Europe, forming glaucous or grey-green, low tussocks on salt-marshes and rocks; it resembles *C. lachenalii* more than any other British species, but has narrower, often channelled leaves, entirely female lateral spikes and concolorous, scarcely beaked utricles.

C. parallela (Laest.) Sommerf. has already been mentioned under the closely allied *C. dioica*; it might be expected to occur in some of the high level flush communities in Scotland.

C. stenolepis Less. is a plant of bogs in N Scandinavia, resembling *C.* × *grahamii* but fertile and often more robust, with dark greenish- or purplish-brown utricles, rather inflated and gradually narrowed into the beak; it may be of similar origin to *C.* × *grahamii* but behaves as a normal fertile species, and might be expected to occur in similar sites.

C. paleacea Schreber ex Wahlenb., assumed to be one of the species from which *C. recta* is derived, is also a plant of estuarine sites and salt-marshes in Scandinavia; it is a very distinctive species with long-pedunculate, pendent, yellowish or pale reddish-brown ♀ spikes and the midrib of the ♀ glumes excurrent as a very long, serrulate awn.

Other possible species include *C. capitata* L., a monoecious species of subgenus *Primocarex* with the spike ♂ above, and 2 stigmas to be looked for in base-rich peaty sites in N Scotland; and also *C. glacialis* Mackenzie, related to *C. ornithopoda*, a densely caespitose plant with suberect stems 3–15 cm, short leaves, solitary ♂ spike, 1–3 small, overlapping ♀ spikes (the upper not overtopping the ♂ spike), dark purplish-brown ♀ glumes and veinless, brownish-green utricles, 2–2.5 mm, abruptly contracted into a short beak. It is a plant of very exposed, dry, stony calcareous places in Iceland and Scandinavia.

REFERENCES AND
SELECTED BIBLIOGRAPHY

BABINGTON, C. C. 1843. *Manual of British Botany*. London.

BENNETT, A. 1897. Notes on British plants, 2. *Carex. J. Bot., Lond.*, **35**: 244–252, 259–264.

BERGGREN, G. 1969. *Atlas of Seeds and Small Fruits of NW European Plant Species with Morphological Descriptions. Pt. 2. Cyperaceae.* Swedish Nat. Sci. Council, Stockholm.

BERNARD, J. M. 1976. The life history and population dynamics of shoots of *Carex rostrata. J. Ecol.*, **64**: 1045–48.

BIRSE, E. L. & ROBERTSON, J. S. 1967. Vegetation in the soils of the country around Haddington and Eyemouth (eds. J. M. Ragg & D. W. Futty). *Mem. Soil Surv. Scot.*, Sheets 33, 34 and 41 (part).

BIRSE, E. L. & ROBERTSON, J. S. 1973. Vegetation in the soils of Carrick and the country around Girvan. *Mem. Soil Surv. Scot.*, Sheets 7 & 8.

BIRSE, E. L. & ROBERTSON, J. S. 1976. *Plant Communities and Soils of the Lowland and Southern Upland Regions of Scotland.* Aberdeen.

CALLAGHAN, T. V. 1976. Growth and population dynamics of *Carex bigelowii* in an alpine environment. Strategies of growth and population dynamics of Tundra plants 3. *Oikos*, **27**: 402–413.

CHATER, A. O. 1980. *Carex* in *Flora Europaea* (ed. T. G. Tutin *et al.*) **5**: 290–323. Cambridge.

CLAPHAM, A. R., TUTIN, T. G. & WARBURG, E. F. 1962. *Flora of the British Isles*, 2nd ed., Cambridge.

COOMBE, D. E. 1954. *Carex humilis* Leyss., pp. 111–113 in C. D. Pigott & S. M. Walters. On the interpretation of the distribution shown by certain British species of open habitat. *J. Ecol.*, **42**: 95–116.

CLYMO, R. 1962. An experimental approach to part of the calcicole problem. *J. Ecol.*, **50**: 707–731.

CRAWFORD, FRANCIS C. 1910. *Anatomy of the British Carices.* Edinburgh.

DAMMAN, A. W. H. 1963. Key to the *Carex* species of Newfoundland by vegetative characteristics. *Dept of Forestry Publication* No. 1017. Ottawa.

DANDY, J. E. 1958. *List of British Vascular Plants*. London.

DAVID, R. W. 1977. The distribution of *Carex montana* L. in Britain. *Watsonia*, 11: 377–378.

DAVID, R. W. 1978a. The distribution of *Carex digitata* L. in Britain. *Watsonia*, 12: 47–49.

DAVID, R. W. 1978b. The distribution of *Carex elongata* L. in the British Isles. *Watsonia*, 12: 158–160.

DAVID, R. W. 1979a. The distribution of *Carex humilis* Leyss. in Britain. *Watsonia*, 12: 257–258.

DAVID, R. W. 1979b. The distribution of *Carex rupestris* All. in Britain. *Watsonia*, 12: 335–337.

DAVID, R. W. 1980a. The distribution of *Carex ornithopoda* Willd. in Britain. *Watsonia*, 13: 53–54.

DAVID, R. W. 1980b. The distribution of *Carex rariflora* (Wahlenb.) Sm. in Britain. *Watsonia*, 13: 124–125.

DAVID, R. W. 1981a. The distribution of *Carex ericetorum* Poll. in Britain. *Watsonia*, 13: 225–226.

DAVID, R. W. 1981b. The distribution of *Carex punctata* Gaudin in Britain, Ireland and the Isle of Man. *Watsonia*, 13: 318–321.

DAVID, R. W. 1982. The distribution of *Carex maritima* Gunn. in Britain. *Watsonia*, (in the press).

DAVIES, E. W. 1953a. Notes on *Carex flava* and its allies. I. A sedge new to the British Isles. *Watsonia*, 3: 66–69.

DAVIES, E. W. 1953b. Ibid. II. *Carex lepidocarpa* in the British Isles. *Watsonia*, 3: 70–73.

DAVIES, E. W. 1953c. Ibid. III. The taxonomy and morphology of the British representatives. *Watsonia*, 3: 74–84.

DAVIES, E. W. 1956. Cytology, evolution and origin of the aneuploid series in the genus *Carex*. *Hereditas*, 42: 349–365.

EDDY, A. & WELCH, D. 1969. The vegetation of Moor House National Nature Reserve. *Vegetatio*, 16: 239–284.

FAULKNER, J. S. 1970. Experimental studies on *Carex* section *Acutae*. D.Phil. thesis, Oxford Univ. (Unpublished).

FAULKNER, J. S. 1972. Chromosome studies on *Carex* section *Acutae* in NW Europe. *Bot. J. Linn. Soc.*, 65: 271–301.

FITTER, A. & SMITH, C. 1979. *A Wood in Ascam: a Study in Wetland Conservation—Askham Bog 1879–1979*. York.

HARBORNE, J. B. 1971. Distribution and taxonomic significance of flavonoids in the leaves of the Cyperaceae. *Phytochemistry*, 10: 1569–1594.

HEGI, G. 1977. *Illustrierte Flora von Mitteleuropa*, ed. 3. *Cyperaceae*. **2(1):** 241–274. Berlin.

HESLOP-HARRISON, J. W. 1941. *Carex bicolor* All., a sedge new to the British Isles. *J. Bot., Lond.*, **79:** 111.

HESLOP-HARRISON, J. W. 1946. Noteworthy sedges from the Inner and Outer Hebrides with an account of two species new to the British Isles. *Trans. Bot. Soc. Edinb.*, **34:** 270–277.

HESLOP-HARRISON, J. 1953. *New Concepts in Flowering Plant Taxonomy*. London.

HJELMQVIST, H. & NYHOLM, E. 1947. Some anatomical species-characteristics in the Scandinavian Carices-Distigmatae. (In Swedish). *Bot. Notiser*, **1947:** 1–31.

HOLDEN, A. V. 1961. Concentration of chloride in fresh waters and rainwater. *Nature, Lond.*, **192:** 961.

HOLDGATE, M. W. 1955. The vegetation of some British upland fens. *J. Ecol.*, **43:** 389–403.

HYLANDER, N. 1966. *Carex* in *Nordisk Kärlväxtflora*, **2:** 42–188. Stockholm.

JERMY, A. C. 1967. *Carex* section *Carex* (=*Acutae* Fr.) *Proc. B.S.B.I.*, **6:** 375–379.

JERMY, A. C., HIBBERD, D. J. & SIMS, P. A. 1978. Brackish and freshwater ecosystems. Chap. 9 in *The Island of Mull: a Survey of its Flora and Environment* (eds. A. C. Jermy & J. A. Crabbe). London.

JERMY, A. C., JAMES, P. W. & EDDY, A. 1978. Terrestrial ecosystems. Chap. 10 in *The Island of Mull: a Survey of its Flora and Environment* (eds. A. C. Jermy & J. A. Crabbe). London.

KERN, J. H. & REICHGELT, T. J. 1954. *Carex* in *Flora Neerlandica*, **1(3)**. Amsterdam.

KERSHAW, K. A. 1962. Quantitative ecological studies from Landmannahellir, Iceland, II. The rhizome behaviour of *Carex bigelowii* and *Calamagrostis neglecta*. *J. Ecol.*, **50:** 171–180.

KOYAMA, T. 1962. Classification of the family Cyperaceae (2). *J. Fac. Sci. Univ. Tokyo; Sect. 3, Bot.*, **8:** 149–278.

KRECZETOWICZ, V. I. 1936. Are the sedges of subgenus *Primocarex* Kük. primitive ? (In Russian.) *J. Bot. U.R.S.S.*, **21:** 395.

KÜKENTHAL, G. 1909. *Das Pflanzenreich*, **4 (20)**. *Cyperaceae: Caricoideae*. Leipzig.

LAMBERT, J. M. 1951. Alluvial stratigraphy and vegetational succession in the region of the Bure Valley Broads, III. Classification, status and distribution of communities. *J. Ecol.*, **39:** 149–170.

Lousley, J. E. 1976. *Flora of Surrey*. Newton Abbot & London.

McVean, D. N. & Ratcliffe, D. A. 1962. *Plant Communities of the Scottish Highlands*. London.

Metcalf, C. R. 1971. *Anatomy of the Monocotyledons. V. Cyperaceae*. Oxford.

Mitchell, J. & Stirling, A. McG. 1980. *Carex elongata* in Scotland. *Glasg. Nat.*, **20**: 65–70.

Neumann, A. 1952. Vorläufiger Bestimmungsschlüssel für Carex-Arten Nordwestdeutschlands in blütenlosen Zustande. *Mitt. Flor.-Soz. Arb. Gemeinschaft*, NF **3**: 44–77.

Nilsson, Ö. & Hjelmqvist, H. 1967. Studies on the nutlet structure of the south Scandinavian species of *Carex. Bot. Notiser*, **120**: 460–485.

Noble, J. C., Bell, A. D. & Harper, J. L. 1979. The population biology of plants with clonal growth. I. The morphology and structural demography of *Carex arenaria. J. Ecol.*, **67**: 983–1008.

Ratcliffe, D. A. 1964a. Mires and bogs. Chap. 10 in *The Vegetation of Scotland* (ed. J. H. Burnett). Edinburgh.

Ratcliffe, D. A. 1964b. Montane mires and bogs. Chap. 15 in *The Vegetation of Scotland* (ed. J. H. Burnett). Edinburgh.

Ratcliffe, D. A. (ed.) 1977. *A Nature Conservation Review*. Vols 1 & 2. Cambridge.

Savile, D. B. O. & Calder, J. A. 1953. Phylogeny of *Carex* in the light of parasitism by the Smut fungi. *Can. J. Bot.*, **31**: 164–174.

Schmid, B. W. 1980. *Carex flava* L. s.1. im Lichte der r-Selektion. *Inaugural Dissertation Univ. Zürich*. Zürich.

Shaver, G. R., Chapin III, F. S. & Billings, W. D. 1979. Ecotypic differentiation in *Carex aquatilis* on ice-wedge polygons in the Alaskan coastal tundra. *J. Ecol.*, **67**: 1025–1046.

Shepherd, G. J. 1975. Experimental taxonomy in the genus *Carex* section *Vesicariae*. Ph.D. thesis, Edinburgh Univ. (Unpublished).

Shepherd, G. J. 1976. The use of anatomical characters in the intrageneric classification of *Carex* (Cyperaceae). *Hoehnea*, **6**: 33–54.

Simpson, N. D. 1960. *A Bibliographical Index of the British Flora*. Bournemouth.

Smith, D. L. 1966. Development of the inflorescence in *Carex. Ann. Bot.*, NS **30**: 475–486.

Smith, D. L. 1967. The experimental control of inflorescence development in *Carex. Ann. Bot.*, NS **31**: 19–30.

SMITH, D. L. 1969. The role of leaves and roots in the control of inflorescence development in *Carex*. *Ann. Bot.*, NS **33**: 505–514.

SMITH, D. L. & FAULKNER, J. S. 1976. The inflorescence of *Carex* and related genera. *Bot. Review*, **42**: 53–81.

SMITH, J. E. 1830. *The English Flora*, ed. 2. **4**: 126. London.

SPENCE, D. H. N. 1964. The macrophytic vegetation of freshwater lochs, swamps and associated fens. Chap. 9 in *The Vegetation of Scotland* (ed. J. H. Burnett). Edinburgh.

TAYLOR, F. J. 1956. *Carex flacca* Schreb. (Biol. Fl. Brit. Isles). *J. Ecol.*, **44**: 281–290.

TOIVONEN, H. 1974. Chromatographic comparison of the species of *Carex* section *Heleonastes* and some *Carex canescens* hybrids in Eastern Fenno-Scandia. *Ann. Bot. Fennici*, **11**: 225–230.

TOIVONEN, H. 1981. Spontaneous *Carex* hybrids of *Heleonastes* and related sections in Fennoscandia. *Acta Bot. Fenn.*, **116**: 1–51.

TOIVONEN, H. & TIMONEN, T. 1976. Perigynium and achene epidermis in some species of *Carex*, subg. *Vignea* (Cyperaceae), studied by scanning electron microscopy. *Ann. Bot. Fennici*, **13**: 49–59.

WALLACE, E. C. 1975. *Carex*. Pp. 513–540 in Stace, C.A., *Hybridization and the Flora of the British Isles*. London.

WALTER, K. S. 1975. A preliminary study of the achene epidermis of certain *Carex* (Cyperaceae) using scanning electron microscopy. *Michigan Botanist*, **14**: 67–72.

WELLS, T. C. E. 1975. The floristic composition of chalk grassland in Wiltshire. In *Suppl. Fl. Wilts.* (ed. L. F. Stearn), 99–125. Devizes.

WHEELER, B. D. 1978. The wetland plant communities of the River Ant valley, Norfolk. *Trans. Norfolk & Norwich Nat. Soc.*, **24**: 153–187.

WHEELER, B. D. 1980a. Plant communities of rich-fen systems in England and Wales. I. Introduction. Tall sedge and reed communities. *J. Ecol.*, **68**: 365–395.

WHEELER, B. D. 1980b. Ibid. II. Communities of calcareous mires. *J. Ecol.*, **68**: 405–420.

WHEELER, B. D. 1980c. Ibid. III. Fen meadow, fen grassland and fen woodland communities, and contact communities. *J. Ecol.*, **68**: 761–788.

WILMOTT, A. J. 1938. *Carex spiculosa* var. *hebridensis* A. Benn. *J. Bot., Lond.*, **76**: 137–141.

VONK, D. H. 1979. Biosystematic studies in the *Carex flava* complex. I. *Acta Bot. Neerl.*, **28**: 1–20.

GLOSSARY

Aerenchymatous (of roots): possessing large, regular air spaces in the ground tissue.

Abaxial: that side facing away from the axis.

Acicular: stiff and pointed, like a needle.

Acuminate: with a long fine point.

Acute: narrowed into a short point.

Adaxial: that side facing towards the axis.

Apiculate: rounded but with a short point.

Apomixis: the production of a fertile seed without the fusion of male and female nuclei.

Arcuate: curved like a bow.

Arista: an awn or bristle.

Aristate: possessing an arista.

Auriculate: furnished with ear-like appendages.

Basifixed (of anthers): having the filament attached at or near the base of the anther.

Caespitose: tufted.

Caducous: dropping off early.

Connate: joined so as not to be separated without tearing.

Contiguous (of spikes): touching each other, end to end or overlapping.

Distant (of spikes): when the distance between spikes is about twice the length of those spikes.

Divaricate: spreading asunder at a wide angle.

Dorsifixed (of anthers): having the filament attached at or near the middle of the back of the anther.

Ecotone: area with vegetation transitional between two or more different plant associations.

Ellipsoid: a solid body with an ellipse-shaped silhouette.

Erose: having an irregularly toothed margin.

Eutrophic: nutrient-rich (usually either from a fertile soil or from being fed by a mineral-rich flow of water); pH usually above 6.

Excurrent (of a vein): running out beyond the margin of a leaf or glume.

Falcate: curved like the blade of a reaper's sickle.

Felty (of roots): having a thick cover of root-hairs.

Fibrilla: thin strands of leaf-sheath tissue usually forming a ladder-like network.

Fibrillose: possessing fibrillae.

Fibrous (of scales and sheaths): the result of decay on the tissue leaving only the vascular bundles (veins).

Filiform: thread-like.

Fimbriate: fringed.

Flaccid: not able to hold up its own weight.

Flexuous: twisted or bent alternately in opposite directions.

Fusiform: swollen in the middle and tapering to each end like a spindle.

Globose: round, like a ball.

Glumaceous: glume-like.

Hyaline: colourless and transparent.

Inflated (of utricles): swollen more than is necessary to hold the nut.

Inrolled: having the margins rolled upwards towards the midrib.

Keeled (of leaves): with a prominent midrib below (see *Fig. 4*).

Lanceolate: shaped like a lance, i.e. broadest below the middle with length to breadth ratio about 3 to 1.

Lingulate: tongue-shaped (see *Fig. 10*).

Lusitanic (of distribution): confined in England and Ireland to the warmer SW parts.

Mesotrophic: neither nutrient-rich nor excessively nutrient-poor; only occasionally flushed with nutrient-rich water.

Midrib: the median vein and, in the case of glumes, the near-by tissue.

Monopodial (of rhizomes): a system in which the apical bud continues to form a creeping stem and upright shoots are formed from lateral buds (see Fig. 2).

Mucro: a stiff, short point.

Mucronate: abruptly terminating in a stiff, short point.

Nerved (of utricles): having conspicuous but not projecting veins.

Ob (prefix): signifying an inversion of the shape, e.g. obovate: egg-shaped with the egg standing on the narrow end.

Obtuse: blunt.

Oceanic (of distribution): confined to areas affected by the Atlantic.

Oligotrophic: nutrient-poor; usually infertile peaty soil with a low pH (usually below 5.5).

Orbicular: circular.

Overwintering (of leaves): remaining green throughout the winter season.

Papillate: with minute protuberances.

Patent: diverging from the axis at almost 90°.

Persistent (of scales and sheaths): not quickly decaying; remaining intact on dying.

Pilose: hairy with distinct long hairs.

Plano-convex (of utricles): more or less flat on the inner (adaxial) side, rounded on the opposite (abaxial) side.

Plicate: folded longitudinally (see *Fig. 7*).

Protandrous: having the pollen maturing before the stigmas are receptive.

Protogynous: having the stigmas receptive before the pollen is ripe.

Pruinose: having a waxy, whitish, powdery 'bloom' (on the surface).

Pyriform: shaped like a pear.

Ramet: fragment of a clone.

Remote (of spikes): when the distance between spikes is at least three times the length of those spikes.

Retuse: with a rounded, shallowly notched end.

Ribbed (of utricles): having pronounced veins near the surface.

Scarious: of a thin, dry, membranous texture and usually colourless.

Septa: cellular crosswalls cutting longitudinal air-tubes of leaf-sheaths into a brickwork-like arrangement.

Setaceous: bristle-like.

Spike: a branch axis bearing flowers.

Spikelets: the ultimate limb of a branched inflorescence (panicle) bearing flowers.

Striate: having fine lines or grooves.

Sub (prefix): somewhat, not completely.

Subulate: awl-shaped (usually with a fine, sharp point).

Sympodial (of rhizomes): a system in which the apical bud has finite growth in the form of an aerial shoot and lateral buds from its base produce further creeping stems (see Fig. 1).

Terete: circular in transverse section.

Tomentose: covered with short, stiff, dense hairs.

Trigonous: having three angles and three faces between them.

Truncate (of beak): ending abruptly as if cut straight across (see *Fig. 31*).

Ventricose: swollen, especially on one side.

Vivipary: the production within the ovary of a seedling without prior fertilisation.

INDEX TO ENGLISH NAMES

Bold face *indicates species number*

261

INDEX TO LATIN NAMES

Names accepted in this book in light face; *synonyms* (used in British floras) *in italics*. The **bold figures** refer to species number and synonyms, not always mentioned in the text, are equated to a species by reference to this number. Those species not described but otherwise mentioned in the text are referred to a page number in light face.

BSBI HANDBOOKS

Each handbook deals in depth with one or more difficult groups of British and Irish plants.

No. 1 SEDGES OF THE BRITISH ISLES
A. C. Jermy, A. O. Chater and R. W. David. 1982. 268 pages, with a line drawing and distribution map for every species. Paperback. ISBN 0 901158 05 4.

No. 2 UMBELLIFERS OF THE BRITISH ISLES
T. G. Tutin. 1980. 197 pages, with a line drawing for each species. Paperback. ISBN 0 901158 02 X. Out of print: new edition in preparation.

No. 3 DOCKS AND KNOTWEEDS OF THE BRITISH ISLES
J. E. Lousley and D. H. Kent. 1981. 205 pages, with many line drawings of native and alien taxa. Paperback. ISBN 0 901158 04 6. Out of print: new edition in preparation.

No. 4 WILLOWS AND POPLARS OF GREAT BRITAIN AND IRELAND
R. D. Meikle. 1984. 198 pages, with 63 line drawings of all species, subspecies, varieties and hybrids. Paperback. ISBN 0 901158 07 0.

No. 5 CHAROPHYTES OF GREAT BRITAIN AND IRELAND
J. A. Moore. 1986. 144 pages, with line drawings of 39 species and 17 distribution maps. Paperback. ISBN 0 901158 16 X.

No. 6 CRUCIFERS OF GREAT BRITAIN AND IRELAND
T. C. G. Rich. 1991. 336 pages, with descriptions of 140 taxa, most illustrated with line drawings and 60 with distribution maps. Paperback. ISBN 0 901158 20 8.

No. 7 ROSES OF GREAT BRITAIN AND IRELAND
G. G. Graham and A. L. Primavesi. 1993. 208 pages, with descriptions and illustrations of 12 native and eight introduced species, and descriptions of 83 hybrids. Distribution maps are included of 31 selected species and hybrids. Paperback. ISBN 0 901158 22 4.

No. 8 PONDWEEDS OF GREAT BRITAIN AND IRELAND
C. D. Preston. 1995. 350 pages. Covers 21 species and 26 hybrids with many full page illustrations, including habit and details of leaves, stems and stipules and up-to-date distribution maps. Includes full descriptions, keys, and introduction with accounts of history, evolution and life histories. Paperback. ISBN 0 901158 24 0.

Available from the official agents for B.S.B.I. Publications:
F. & M. Perring, Green Acre, Wood Lane, Oundle, Peterborough, PE8 4JQ, England.
Tel: 01832 273388   Fax: 01832 274568

BOTANICAL SOCIETY OF THE BRITISH ISLES (B.S.B.I.)

BSBI

The B.S.B.I. was founded in 1836 and has a membership of 2,700. It is the major source of information on the status and distribution of British and Irish flowering plants and ferns. This information, which is gathered through a network of county recorders, is vital to their conservation and is the basis of the *Red Data Books* for vascular plants in Great Britain and Ireland. The Society arranges conferences and field meetings throughout the British Isles and, occasionally, abroad. It organises plant distribution surveys and publishes plant atlases and handbooks on difficult groups such as sedges and willows. It has a panel of referees available to members to name problem plants. Through its Conservation Committee it plays an active part in the protection of our threatened plants. It welcomes all botanists, professional and amateur alike, as members.

Details of membership and any other information about the Society may be obtained from:

The Hon. General Secretary,
Botanical Society of the British Isles,
c/o Department of Botany,
The Natural History Museum,
Cromwell Road,
London, SW7 5BD.